AS UNIT 2

STUDENT GUIDE

CCEA

Geography
Human geography

Tim Manson

HODDER
EDUCATION
AN HACHETTE UK COMPANY

This Guide has been written specifically to support students preparing for the CCEA AS Unit 2 Geography examinations. The content has been neither approved nor endorsed by CCEA and remains the sole responsibility of the author.

Orders: please contact Hachette UK Distribution, Hely Hutchinson Centre, Milton Road, Didcot, Oxfordshire, OX11 7HH. Telephone: +44 (0)1235 827827. Email: education@hachette.co.uk. Lines are open from 9 a.m. to 5 p.m., Monday to Friday. You can also order through our website: www.hoddereducation.co.uk.

© Tim Manson 2016

ISBN 978-1-4718-6412-4

First published in 2020 by
Hodder Education,
An Hachette UK Company
Carmelite House
50 Victoria Embankment
London EC4Y 0DZ
www.hoddereducation.co.uk

First printed 2016

Impression number 7

Year 2022

Cover photo: Alistair Hamill

Typeset by Integra Software Services Pvt Ltd, Pondicherry, India

Printed and bound by CPI Group (UK) Ltd, Croydon, CR0 4YY

Hachette UK's policy is to use papers that are natural, renewable and recyclable products and made from wood grown in well-managed forests and other controlled sources. The logging and manufacturing processes are expected to conform to the environmental regulations of the country of origin.

MIX
Paper | Supporting responsible forestry
FSC™ C104740

Contents

Content Guidance

Questions & Answers

■ Getting the most from this book

Exam tips

Advice on key points in the text to help you learn and recall content, avoid pitfalls, and polish your exam technique in order to boost your grade.

Knowledge check

Rapid-fire questions throughout the Content Guidance section to check your understanding.

Knowledge check answers

1 Turn to the back of the book for the Knowledge check answers.

Summaries

■ Each core topic is rounded off by a bullet-list summary for quick-check reference of what you need to know.

Exam-style questions

Commentary on the questions

Tips on what you need to do to gain full marks, indicated by the icon (e)

Sample student answers

Practise the questions, then look at the student answers that follow.

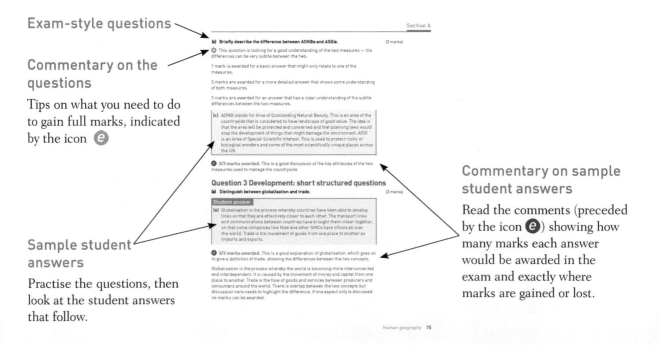

Commentary on sample student answers

Read the comments (preceded by the icon (e)) showing how many marks each answer would be awarded in the exam and exactly where marks are gained or lost.

■ About this book

Much of the knowledge and understanding needed for AS geography builds on what you have learned for GCSE geography, but with an added focus on geographical skills and techniques, and concepts. This guide offers advice for the effective revision of **AS Unit 2: Human geography**, which all students need to complete.

The AS 2 external exam paper tests your knowledge and application of aspects of human geography with a particular focus on population, settlements and the geography of development. The exam lasts 1 hour 15 minutes. The unit makes up 40% of the AS award or 16% of the final A-level grade.

To be successful in this unit you have to understand:

- the key ideas of the content;
- the nature of the assessment material — by reviewing and practising sample structured questions; and
- how to achieve a high level of performance within the examination.

This guide has two sections:

Content guidance — this section summarises some of the key information that you need to know to be able to answer the examination questions with a high degree of accuracy and depth. In particular, the meaning of key terms is made clear and some attention is paid to providing details of case study material to help to meet the spatial context requirement within the specification. Students will also benefit from noting the **Exam tips** that will provide further help in determining how to learn key aspects of the course. **Knowledge check questions** are designed to help learners to check their depth of knowledge — why not get someone else to ask you these?

Questions & Answers — this section includes several sample questions similar in style to those you might expect in the exam. There are some sample student responses to these questions as well as detailed analysis, which will give further guidance in relation to what exam markers are looking for to award top marks.

The best way to use this book is to read through the relevant topic area first before practising the questions. Only refer to the answers and examiner comments after you have attempted the questions.

Content Guidance

■ Topic 1 Population

Population data

National census taking

The Office of National Statistics (2010) notes the following:

> A census is a count of the population. We have to have one in the UK every 10 years to find out more about who we are as a nation. We ask everyone to tell us a little bit about themselves to help census users decide how best to plan, fund and deliver the everyday services we all need — like housing, education, healthcare and transport.

Brief history of the census

Within the UK, a full census has been taken every ten years since 1801 (apart from 1941). The amount of information collected has increased over the years, with many of the questions providing information on the population trends, resources and requirements for public services in an area. The census is seen as a **static** measure, as it takes a snapshot of the population of a country on a particular day.

The census in MEDCs

The most recent census in the UK (an example of a more economically developed country, or MEDC) took place on Sunday 27 March 2011. This information can support government services in the following ways.

- Population: knowing how many people are living in an area helps the government allocate funding where it is needed.
- Education: the census helps to plan the location and changes needed to education services required in the future.
- Health and disability: health services can be planned to make sure that healthcare is concentrated in the areas where it will be most needed.
- Housing: housing needs can be better planned if the authorities know what the demand is now and what it will be in the future.
- Employment: information can be used to help plan jobs and training needs.
- Ethnic groups: census information can be used to help allocate resources and ensure that all groups are treated fairly.
- Transport: working out how and where people travel to work can help the government to understand pressures on our transport system and improve public transport.

> **Knowledge check 1**
>
> Why is the census useful to national governments?

The census in LEDCs

In recent years, the United Nations (UN) has supported a large number of less economically developed countries (LEDCs) in improving their demographic data collection, but there can still be issues regarding the management and quality of data collection.

Data collection and reliability issues

Collecting the information across countless small villages and towns can be difficult and can compromise the accuracy of the data.

- Literacy levels: poor education rates in many LEDCs mean that few people can read and write, and many would be unable to complete a written census form.
- Lifestyles: nomadic tribes and families can be difficult to find and might migrate across international boundaries. In Kenya, the logistics for collecting information from the 'pastoral communities' require more resources.
- Size: large countries with countless scattered villages (e.g. India) can make the process of census organisation difficult.
- Fluid population: centres of mass migratory population, like the shanty towns in São Paulo, Brazil, have a transient population, which is difficult to analyse.
- Cost: many LEDCs do not have the money to spend on such a task.
- Mapping inaccuracies: households might be left out if the mapping of areas is incomplete.
- Transport difficulties: it can be difficult for enumerators to gain access to some places, which can be made worse by seasonal rains or weather patterns.
- Cultural traits: in some areas of the Middle East, male enumerators are not permitted to interview women.
- Language barriers: tribal/ethnic languages can sometimes cause obstacles. In Cameroon, for example, there are more than 30 different language groups.
- Lack of reference: some people are not aware of what age they are, as they have no real point of reference from which to work this out.

Vital registration

Vital registration is a more **dynamic** aspect of population data, where the number of births and deaths within a country is monitored every day and is always changing. People are required to register births, still-births, deaths and marriages. This has been compulsory in the UK since 1847.

Registration of births

Usually a birth in Northern Ireland needs to be registered within 42 days. Often information is collected within the maternity ward of the hospital and parents can then pick up a birth or adoption certificate from the General Register Office for Northern Ireland (GRONI).

Registration of deaths

A death should be registered as soon as possible to allow funeral arrangements to go ahead and no later than 5 days from the date of occurrence except where the matter

> **Exam tip**
>
> Make sure that you understand the key issues behind census reliability in LEDCs and MEDCs. Issues with data collection practices in LEDCs are more obvious than in MEDCs, but questions on reliability issues can also be asked in relation to MEDCs.

has to be dealt with by a coroner. The person registering the death must go to the registrar with the medical certificate showing the cause of death.

Vital registration measures are extremely reliable in MEDCs and within LEDCs. Most LEDCs have some form of mechanism for registering the birth of a baby, however, there are some places where a written record of births is still not kept. There can be falsification claims for both births and deaths. This is more likely for deaths as associated family members might have more to gain from getting a certificate to say that a family member is deceased.

Knowledge check 2

What might be a potential issue for the reliability of vital registration in an MEDC?

Case study

MEDC: population data in Northern Ireland (part of the UK)

How census data are collected

The census is a paper questionnaire that is delivered by post to each household across the UK. The Northern Ireland Statistics and Research Agency (NISRA) notes that: 'Everyone was asked the same questions on the same day so that we could get a snapshot of the population. This information is used to estimate the number of people and households in each area, and their characteristics.' A team of 'enumerators' then collects the completed surveys in the weeks that follow before the answers are scanned into computers using optical readers. However, for the first time in 2011, householders were given a code that gave them the option of completing their census form online. Over 94% of respondents returned their census form and a high level of quality assurance checking was carried out to ensure the validity of the completed forms.

The most recent census in the UK took place on Sunday 27 March 2011. NISRA organised the census in Northern Ireland.

A Census Advisory Group begins to plan each ten-year census many years in advance. Legislation is required in the UK Parliament and the Northern Ireland assembly to allow the census to take place. A variety of surveys will then take place to make sure that the quality of the census is assured and that the census is harmonised with the census held in the other regions of the UK. A small pilot study, or Census Rehearsal, is usually carried out 2 years before the main event.

The 2011 census in Northern Ireland is estimated to have cost around £22 million between 2008 and 2014. The 2011 census for England and Wales is estimated to have cost £482 million over the same period.

The reliability of census data

The UK census is one of the most reliable in the world. However, issues can affect the reliability of the results.

- Government interference: some people, e.g. those that might be claiming unemployment benefit while also working, might want to avoid the government knowing they are in employment and give false information.
- Confidentiality issues: some people do not want the government to have information on file about them and therefore refuse to fill in the form.
- Language, sight and special needs: some people find it difficult to read or to understand the questions on the form and might need help.

Questions on the UK census

The most important feature of the census is the range of questions that it asks.

Demographic questions: how many people live in your household? What gender the people who live here? How long have you lived in the UK? Were you born outside the UK?

Social questions: what passport do you hold? How would you describe your national identity? What is your ethnic group? What religion do you belong to? What is your main language?

Economic questions: how many cars or vans are owned, or available for use, by members of this household? What qualifications do you have? What is your main job? How many hours a week do you usually work? How do you usually travel to your main place of work?

Use made of the census data

Much of the information collected from the census is released through the NISRA website. There is a wide range of formats available to present and analyse the census information.

- **Standard output tables:** A wide selection of tables are released in order to share the information the census teams collect.
- **Geographical Information Systems** (GIS) information: NISRA releases some of the census information in a GIS format, called NINIS. Usually this information takes a few years to process and be uploaded to the NISRA website. This information can be viewed at a wide range of scales and outputs including area maps and interactive content.

- **Forward planning:** The Northern Ireland census provides information so that the government can plan for future years. Important decisions can be made as to what services should be funded in the future.

Vital registration in Northern Ireland

In Northern Ireland births are registered within 42 days. In most cases the birth details are collected within 24 hours from the maternity ward where a baby is born. The record of birth is called a 'birth certificate' and this is usually required in order to allow a child to engage with particular government services like free healthcare and education. Parents can collect the birth certificate from local registrar general offices, which are usually found within local council offices. Deaths are registered very quickly and this must be done before funeral arrangements can be carried out. The legal implications for both certificates mean that this method of collection is very reliable and there are few births or deaths that go unrecorded. The only exception is that some children born at home rather than in hospital may be missed out.

Knowledge check 3

What are the main issues of reliability for an MEDC census?

Case study

LEDC: population data in Kenya

How census data are collected

In a recent census carried out in Kenya on 24–5 August 2009, enumerators were trained to visit precise areas and count the number of people living in each. The census included questions on age, sex, tribe or nationality, religion, birth place, orphanhood, number of children, job, number of livestock, main source of water, main type of cooking fuel and access to different aspects of ICT (radio, television, mobile phone, landline and computer).

The Kenya National Bureau of Statistics notes that a census is 'the total process of collecting, compiling, evaluating, analysing and publishing or otherwise disseminating demographic, economic and social data pertaining, at a specified time, to all persons in a country.'

The 2009 census cost 8.4 billion Kenyan Shillings (around £58 million). Around 26,000 supervisors, 112,000 enumerators and over 100,000 village elders were involved in collecting the 12 million questionnaires.

The reliability of census data

Data collection took place in the week beginning 24 August 2009. In an effort to improve reliability, the Kenyan president designated the first day of the census (25 August) as a census/public holiday.

There were a number of constraints and challenges that affected the collection of information in Kenya.

- Delayed payment: there was unrest among some of the census enumerators as there was confusion regarding the employment and payment of some of the workers. Some did not receive the money they had earned as they did not have proper bank accounts and this interfered with the pace and detail collected throughout the process.
- Nomadic/pastoral population: there were challenges in collecting detailed information across the big distances and difficult terrain. Drought conditions in some parts of the country also meant that there was more nomadic movement than usual, some across international boundaries.
- Ethnic tensions: there was some sensitivity in relation to some of the ethnicity (tribe) questions. In some cases people were reluctant to reveal their ethnic background.

Questions on the Kenyan census

The main method of collection was a household survey where all the people who had spent the census night were included. The questions tried to generate information on the sex, age, marital status and place of birth for each person. In addition the census asked questions on disability, the use of ICT, household amenities, assets, livestock and recent births and deaths in the household. A very important question asked people about their ethnic background.

Use made of census data

The Kenyan National Bureau of Statistics notes that the information gathered by the census, 'helps determine locations for schools, roads, hospitals and more…it provides all levels of governments, business, industry, media, academia and independent organisations with social, economic and demographic information needed for making decisions.'

Businesses can also use the information to help them locate services and housing. The 2009 census in Kenya was also going to help redraw some of the election boundaries for parliamentary elections.

The results were collated so that the government would be able to better plan and manage the economic and social needs of the country. Detailed census results have been made available for download from the Bureau of Statistics website.

Vital registration in Kenya

The reliability of vital registration data has improved within many LEDCs in recent years. In Kenya a campaign was launched in 2005 to improve the reliability of vital statistics. Up to that point only 34% of the rural population and 84% of the urban population were registered. A universal birth and death registration programme has now been implemented where parents must complete a notification form at the tribal chief's office when a child is born — but this process can still take up to 2 years.

Exam tip

Make sure that you are able to compare the reliability of both of the case study countries and be able to compare the different aspects of each.

Summary

- A census is a survey of the population. In most countries this is carried out as a written or online questionnaire/form that heads of household must complete.
- In the majority of countries a census is carried out every 10 years.
- A census is a static measure of population as it takes a snapshot of the population make-up of a country.
- Each national census has questions that are specific to the particular needs of that country. In Northern Ireland, the census questions can be divided up into three distinct categories: demographic, social and economic.
- A census taken in an MEDC is likely to be at least 99.9% accurate. However, there can still be issues related to governmental interference, confidentiality and access to the census form.
- Vital registration is a dynamic aspect of population data where the numbers of births and deaths in a country are monitored.
- In most MEDCs vital registration is carried out at a local government level and reliability is close to 100%.
- In LEDCs there are issues of reliability as many countries have yet to develop universal registration practices and laws. In general, the more rural the community, the more difficult it can be to register a birth.
- The last census to take place in Northern Ireland was taken on 27 March 2011. This census was at least 94% reliable and included some demographic, social and economic questions. The census data were released using the NISRA website over a number of years and a variety of outputs including the use of interactive GIS maps. The collection of vital registration in Northern Ireland is extremely accurate.
- The last census to take place in Kenya was taken on 24–25 August 2009. The census was carried out with a great deal of accuracy but was not as accurate as that in Northern Ireland. Over 12 million census forms were collected. There were more reliability issues with the Kenyan census including confusion with the delayed payment of staff, ethnic tensions and issues in locating and questioning any nomadic/pastoral people. The collection of vital registration in Kenya is relatively recent and much of the population still goes unrecorded.

Population change

Fertility and mortality measures

The population balance

The population of a country is dynamic and always changing:

births ± deaths = natural population change

Typically crude birth rates (CBR) are high in LEDCs (30–40 per 1,000) and low in MEDCs (8–15). Also typically, crude death rates (CDR) are low in MEDCs (6–14) while in LEDCs some remain high (> 30) and others have fallen quickly (to around 8–10).

When there is a growth of population (a higher birth rate than death rate) we say that there is a **natural increase**. If there is a decline in the number of people (a lower birth rate than death rate) we say that there is a **natural decrease**.

Some additional fertility and mortality measures of a population include:

- Total fertility rates (TFR) are usually lower in MEDCs than LEDCs. The global average TFR (according to the UN) is currently 2.36 and has been rapidly falling (from a high of 4.95 in 1955).

Crude birth rate The number of live births each year per 1,000 of the population in an area.

Crude death rate The number of deaths each year per 1,000 of the population in an area.

Total fertility rate The number of children a woman is expected to have during her reproductive lifetime (15–49) based on the current birth rates.

- In 1998 the UN noted that a country needed a TFR of 2.1 to replace its population: the **replacement rate**.
- The **infant mortality rate** (IMR) is often used as a measure of development for a country as it indicates how much money a country spends on healthcare for young children.

Replacement rate
When there are enough children born to balance the number of people who have died.

Infant mortality rate
The number of deaths per 1,000 children (infants) born, in the first year of their life in a given year.

The Demographic Transition Model

The Demographic Transition Model (DTM) is one of the best ways to show how the population of an area can change over time. The model takes countries through four or five stages of development and can help with any explanation for changes in birth and death rate. It is originally based on the work of US demographer Warren Thompson in 1929 but it has gone through many amendments and changes since.

The model was based on the population journey taken by a number of industrialised countries and argues that all countries will pass through a series of changes or stages.

Exam tip

It is really important that you understand fully and are able to define the different elements of population balance, and the key fertility and mortality measures. This is a common short question. Learn the definitions in detail.

Description of the Demographic Transition Model

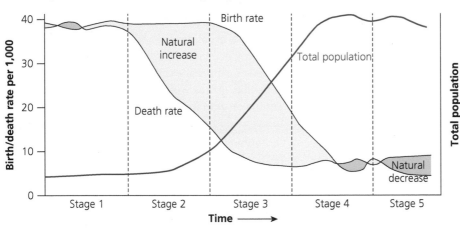

Figure 1 The Demographic Transition Model

The DTM (see Figure 1 and Table 1) is a graph that contains three lines with information. The main aspects of the graph are the two lines that indicate the birth and death rate for the country. The model shows how these rates will change and reduce over time. The third optional line on the graph shows the affect that these changes to births and deaths will have on the total population of the country.

In **Stage 1** the birth and death rates are both high and fluctuating. Both rates will be above 35 per 1,000. The population will be undergoing very fast change as people will not be living long — **life expectancies** (the number of years that someone is expected to live to) will be short. The population will remain small.

In **Stage 2** the main story is that the death rate will begin to fall slowly over a long period of time. As the death rate falls, the birth rate will remain constantly high and this means that the population will begin to increase.

In **Stage 3** the death rate has now fallen to relatively low levels (below 10 per 1,000) but now it is the turn of the birth rate to fall. Birth rate usually falls much faster than death rate. The natural increase of the population will still be evident at this stage but as the birth rate falls, the speed of growth of the population will start to slow down.

In **Stage 4** the birth and death rates are both low (below 10 per 1,000) and will fluctuate to some extent. The total population will remain relatively high.

In **Stage 5** the birth rate (7 per 1,000) will fall below the death rate (9 per 1,000). This means that there are not enough babies being born to replace the population and the size of the total population will be reduced.

Table 1 Stages of development

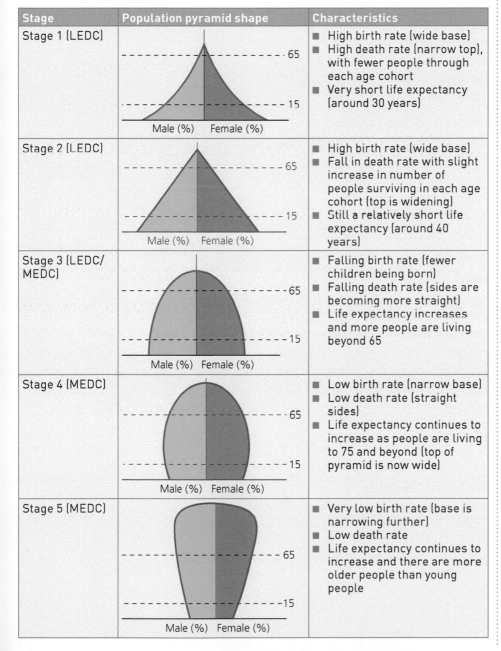

Stage	Population pyramid shape	Characteristics
Stage 1 (LEDC)	Male (%) Female (%)	■ High birth rate (wide base) ■ High death rate (narrow top), with fewer people through each age cohort ■ Very short life expectancy (around 30 years)
Stage 2 (LEDC)	Male (%) Female (%)	■ High birth rate (wide base) ■ Fall in death rate with slight increase in number of people surviving in each age cohort (top is widening) ■ Still a relatively short life expectancy (around 40 years)
Stage 3 (LEDC/MEDC)	Male (%) Female (%)	■ Falling birth rate (fewer children being born) ■ Falling death rate (sides are becoming more straight) ■ Life expectancy increases and more people are living beyond 65
Stage 4 (MEDC)	Male (%) Female (%)	■ Low birth rate (narrow base) ■ Low death rate (straight sides) ■ Life expectancy continues to increase as people are living to 75 and beyond (top of pyramid is now wide)
Stage 5 (MEDC)	Male (%) Female (%)	■ Very low birth rate (base is narrowing further) ■ Low death rate ■ Life expectancy continues to increase and there are more older people than young people

Exam tip

Make sure that you understand the difference between describing the graph used in the DTM and explaining the reasons for the changes between each stage.

Knowledge check 4

Describe the key aspects of the DTM.

Explanation of the Demographic Transition Model

Stage 1: 1750–1800 (Great Britain) or very poor LEDCs

■ Birth rates and death rates are high but fluctuating, giving a small population growth.
■ It is a youthful population structure.
■ Birth rates are high because of:
  – a lack of family planning,
  – a high infant mortality rate — parents have large families to ensure that some children reach adulthood and
  – children are needed to work the land.
■ Death rates are high (especially among children) owing to:
  – the prevalence of disease/famine across the land,
  – poor standards of living and hygiene and
  – basic or non-existent medical care.

Stage 2: 1800–80 (Great Britain) or LEDCs

■ Birth rates remain high as the food supply becomes more reliable.
■ Death rates fall dramatically because:
  – mortality crises like plague or famine have been eliminated,
  – of improvements in nutrition and standards of living,
  – of improvements in medical care and
  – infant mortality rates decrease due to improved healthcare.

Stage 3: 1880–1950 (Great Britain) or NICs/poorer MEDCs

■ Death rates continue to fall and stabilise.
■ The average woman is having 5.5 children in 1871 but this has fallen to 2.4 children in 1921. Birth rates fall quickly due to a number of factors:
  – marriage is delayed or traditional methods of birth control are developed and used,
  – lower infant mortality rate means less need to continue to have children,
  – increasing industrialisation means that fewer workers are needed in urban factories,
  – there is increased desire for material possessions,
  – improved roles and equality for women means more women enter the workplace.

Stage 4: 1950 to present (Great Britain) or MEDCs

■ Birth and death rates become low but fluctuating.
■ The rate of population growth in Great Britain slows down quickly so that by 1960 the population is only growing by 5% each year and by the 1980s this has slowed to 1.9%.
■ Demands for labour mean that Great Britain has to search for migrant workers to keep the economy moving.

Stage 5: a few MEDCs

- Birth rate falls below the death rate, which means that there are not enough babies being born to replace the population (e.g. Italy, Russia).
- People in these countries are more concerned with jobs/careers and earning money than in settling down to raise a family. Children are seen as a draw on resources rather than an asset.

Table 2 Key facts relating to the demographic transition for Great Britain, 1800–2015

Stage/period	Stage 1: 1750–1800	Stage 2: 1800–80	Stage 3: 1880–1950	Stage 4: 1950–present
Birth rate	High (37/1,000)	High (30/1,000)	Down (16/1,000)	Low (13/1,000)
Death rate	High (31/1,000)	Down (19/1,000)	Low (13/1,000)	Low (9/1,000)
Life expectancy	Low (45)	Improving (55)	Improving (65)	High (75)
Population (millions)	10.5	29	43	64

Limitations of the Demographic Transition Model

The DTM cannot be easily applied to every population change. In many LEDCs, recently there has been a huge fall in the death rate and the birth rate, which is starting to make the demographic distinction between MEDC and LEDC more difficult.

- The model is eurocentric: it is based on the journey that many European MEDCs have gone through, yet many of the LEDC countries have been moving through the stages at a much faster rate.
- The model assumes that death rates in Stage 2 fall due to industrialisation. This is not the experience in many LEDCs. Often other reasons such as political intervention, economic investment by other richer countries or aid/relief organisations have had an effect on both death and birth rates.
- R. Woods wrote that 'the theory lacks rigour; it has logical inconsistencies; it involves a hotchpotch of causal variables and cannot be expected to have universal applicability.'

The Epidemiological Transition

Description of the Epidemiological Transition

The Epidemiological Transition is a theory put forward by Abel Omran in 1971, which allows geographers to study how countries move over time from being dominated by infectious diseases to being dominated by chronic diseases. This is mostly due to the continued improvement and expansion of public health and sanitation systems. The model has some similarities to the DTM but it focuses on causes of death and shows how the causes should be changing as the country becomes more developed.

Exam tip

There are many different types of question that could be asked in reference to demographic transition. Make sure that you have a clear understanding of the theory and the practical applications of this theory to the journey that particular countries may go through.

Knowledge check 5

Explain how the demographic transition experience for Great Britain helps to prove the reliability of the DTM.

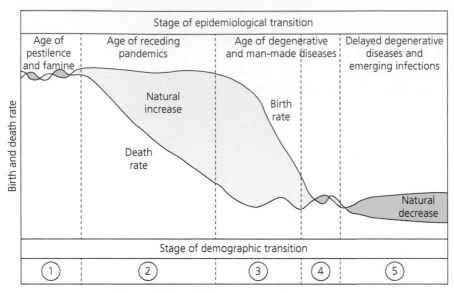

Figure 2 The Epidemiological Transition

Figure 2 shows the three main stages of the epidemiological transition of mortality.

Stage 1: the age of pestilence and famine

Death rates remain very high in this stage, as the causes of death are mainly infectious diseases such as chicken pox, small pox, measles, tuberculosis and influenza. Sometimes these are referred to as being **exogenetic** (as they are not linked to genetic factors). Life expectancies are very low (between 20 and 30 years).

Stage 2: the age of receding pandemics

Death rates begin to fall as developments in medical science and sanitation allow infectious diseases to be controlled. Life expectancies increase from 30 to 50 years.

Stage 3: the age of degenerative and man-made diseases

As life expectancies continue to increase and mortality rates decline, the population approaches stability at a low level. Modern healthcare, antibiotics and improvements to infant mortality rates combine to allow people to live longer. The cause of death becomes degenerative or **endogenetic** diseases (e.g. Alzheimer's, heart disease, strokes and cancer), which affect people only in later life (Figure 3).

Typically, the stage that a country finds itself at along the epidemiological transition will depend upon its stage of development. LEDCs are more likely to be at either Stage 1 or Stage 2 with high numbers of people still dying from infectious diseases. Towards the end of Stage 2 and into Stage 3, as life expectancies rise, countries are more likely to be MEDCs, where increased spending on health and medical care can make a big difference.

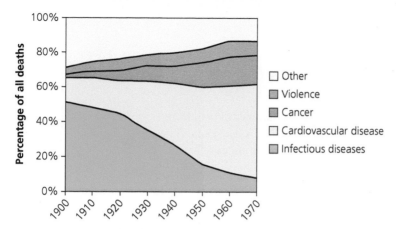

Figure 3 The health transition for mortality in the United States from 1900 to 1970

Limitations of the Epidemiological Transition model

Although the Epidemiological Transition model shows the changes in the cause of death for a particular country, some question whether this change really took place during the 20 century. The change is really just a replacement of infectious diseases by chronic diseases. There is no single explanation for this change, and often the reasons for such change lie in transitions in society, sanitation, access to medicine or improved diagnosis of illness.

Over recent years, the mortality rate in LEDCs has fallen rapidly due to the intervention of aid agencies and intergovernmental organisations like the UN. This change has occurred at a much faster rate than the MEDCs experienced. As a result, many LEDCs now have death rates lower than MEDCs (e.g. Kenya is 9 per 1,000), although as this youthful population gets older it is expected that the death rate will increase again.

Summary

- Population increase occurs when birth rates are higher than death rates.
- Population decrease occurs when death rates are higher than birth rates.
- Birth rates can be measured using the crude birth rate and the total fertility rate.
- Death rates can be measured using the crude death rate and the infant mortality rate.
- The DTM is a graph containing information on birth rates, death rates and the total population of a country. It can be used to monitor the changing population over time for a country.
- Each stage of the DTM indicates the falling death rate and then the falling birth rate of the country, and how this can rapidly increase the population.

- Many of the individual changes to death rate and birth rate are down to social, economic and political change within the population.
- The DTM was originally based on the experience of a number of European countries and there are therefore limitations in applying this model to the demographic experience of many LEDCs.
- The Epidemiological Transition describes the changes in the cause of death in countries over time. Many countries move from the 'age of pestilence and famine', in which infectious diseases cause most deaths, to a more advanced 'age of degenerative and man-made diseases', in which people live longer lives and fall to more genetic diseases like heart disease and cancer.

Population and resources

Underpopulation, overpopulation and optimum population

There have been concerns over the balance between the number of people and the amount of resources available to support these people for a long time. As world population continues to increase, this can put pressure on resources.

Resources are defined as being any aspects of the environment that are used to support human needs.

The **optimum population** of an area is the number of people which, when working with all of the available resources, will return the highest standard of living and quality of life. If the size of the population increases or decreases from the optimum, the standard of living will fall.

The standard of living is usually noted as the interaction between physical and human resources:

$$\text{Standard of living} = \frac{\text{Natural resources (minerals, energy, soils, etc.) x Technology}}{\text{Population}}$$

If the population of the UK was to be reduced to 30 million people this would lead to the abandonment of many resources and a reduction in public services; this would cause a decline in living standards. If the population was to increase to 90 million people this would lead to a serious stress on resources and public services, which would also lead to a decline in living standards.

Overpopulation is when there are too many people relative to the resources and technology available to maintain an adequate standard of living. Countries like Bangladesh and Ethiopia are said to be overpopulated as they might have insufficient food, minerals, water supply or energy resources, and they might suffer from natural disasters such as drought or famine.

Underpopulation is when there are far more resources available in an area than the number of people living there. There might be high levels of food production, energy or minerals that the people living in the area can use. Countries like Canada and Australia export large amounts of food and energy yet the people there have high incomes, very good living conditions and high levels of employment. If the population was to further increase, the standards of living could further improve.

Carrying capacity is the largest population that the environment of an area can actually support with reference to the resources available.

Population sustainability theories of Malthus and Boserup

Organisations such as World Population Balance argue that global footprint information suggests that the world population is already at an unsustainable level. It argues that the population currently uses an equivalent of 1.6 Planet Earths. In addition, they note that if everyone lived to a European standard of living, the world could only support a population of 2 billion people compared to 7.4 billion in 2016.

> **Exam tip**
>
> It is useful to make sure that you can describe the differences between overpopulation and underpopulation compared with the optimum population. Make sure that you can give examples of countries in each case.

> **Knowledge check 7**
>
> What are the main differences between optimum, overpopulation and underpopulation?

Questions about the sustainability of the population have been around for many years. Sustainable development is 'development that meets the needs of the present without compromising the ability of future generations to meet their own needs'. Sustainability is usually identified using the interaction between the 'three pillars' of the economy, society and the environment. A sustainable population is one that can be maintained at the number of people without impacting the quality of life or standards of living of that population (Figure 4).

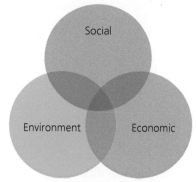

Figure 4 Population sustainability

The idea that a country could become overpopulated was initially raised in the writings of Thomas Malthus. His essay *The Principle of Population* was published in 1798. The core of Malthus' theory is his statement that 'The power of population is indefinitely greater than the power in the Earth to produce subsistence for man. Population, when unchecked, increases in a geometrical ratio. Subsistence increases only in an arithmetical ratio'. Malthus claimed that this would create a 'crisis point' as there would be a finite number of people that a country could actually sustain. At this point a variety of checks would come into play.

Malthus is recognised as having a pessimistic view of the dangers of overpopulation and claimed that food supply was the main limit to population growth. During the time that Malthus was writing, food supply issues and inadequate resources and clothing were common features of everyday life. English towns were growing rapidly due to the emergence of the Industrial Revolution and industrial accidents, pollution, disease and malnutrition complicated urban life. Malthus noted three types of **preventive** or **negative checks**, which were nature's way of controlling the population increase (by controlling birth rate).

- Misery: the impact of disease, famine and war and 'all of the causes which shorten the duration of human life'.
- Vice: Malthus warned against the dangers of practising any kind of 'family planning' and believed that this would only lead to promiscuity.
- Moral restraint: Malthus argued that delayed marriage and abstinence from sexual relations within marriage were highly advisable.

Malthus also noted that a **positive check** could increase mortality (through controlling death rate), e.g. low standards of living/unhealthy conditions that could result in disease, famine and even war.

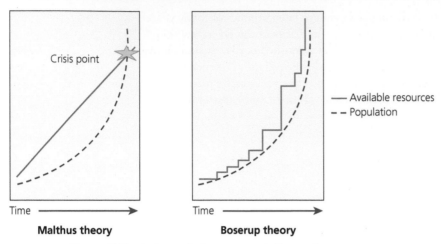

Figure 5 The balance between population and resources

More recently, in 1965, Ester Boserup introduced a more optimistic way of looking at the balance between population and resources in her essay 'The Conditions of Agricultural Growth'. Her starting point was that much of the misery that Malthus outlined had not come to pass, but had instead been a catalyst for innovation. Boserup claimed that the pressure of population growth stimulated human innovation and noted that 'necessity is the mother of invention'. She believed that the population would grow until it reached the carrying capacity and then human innovation would respond to the challenge — people would develop substitutes or improve agricultural technology. Some geographers point to the innovations that came as the result of conflict (following the Second World War) or from food shortages in India that led to the Green Revolution, as good examples of this approach.

Malthus wrote his ideas at the end of the 18th century, yet they still strike a chord with people today. Some recent theorists share his pessimism — they are sometimes called Neo-Malthusians. For example, in his book *The Population Bomb* (1968), Paul Ehrlich supported Malthus' view and noted that 'Each year, food production in the undeveloped countries falls a bit further behind burgeoning population growth, and people go to bed a little bit hungrier.'

Supporters note that the world is now at maximum capacity for agricultural land and fresh drinking water. Famines, food crises, environmental issues, deforestation, floods, natural disasters, soil erosion, crop failure, desertification and to some extent climate change are all examples of Malthusian checks on the population.

However, some argue that Malthus did not make enough allowance for technological change, the development of new 'industrial' agricultural practices or irrigation methods. He also failed to appreciate that the birth rate in countries would likely fall when a certain level of advance development had been reached.

In his book *The Ultimate Resource* (1981), Julian Simon echoes some of Boserup's ideas. He thought that a moderate population growth may have a positive effect in nudging on development, and saw sudden increases in world population as responding to major improvements or inventions.

Evaluating the theories

Table 3 Evaluating the theories of Malthus and Boserup

Malthus	Boserup
Positives	
■ Overpopulation and lack of resources caused food shortages/famines in Ireland/Africa/China.	■ The Green Revolution in the 1950s led the technological revolution in agriculture and produced new high-yielding crops that were more efficient.
■ Soil erosion and desertification in Sub-Saharan Africa is due to overgrazing.	■ New technology/inventions have increased the capacity and effectiveness of agriculture.
■ Conflict in some parts of the world is over resources (water/gold/diamonds).	■ New global charity organisations are investing in global investment programmes to support agriculture.
Negatives	
■ The theory is too simplistic. Population sustainability is more complicated.	■ The theory is based on a 'closed' community, the existence of which in the global community is very unlikely.
■ Globalisation further complicates things, as richer countries import food to support their lack of agricultural produce.	■ International migration further complicates things, as at times of crisis people move and relieve the pressure on the source area.
■ Many countries actually have surplus food rather than a deficit. The wider issue is the fair distribution of resources.	■ Some areas will never be able to sustain people — marginal areas like semi-arid desert can never be improved and in some cases desertification is increasing the amount of land that cannot be farmed.

Exam tip

Examiners like to get students to evaluate the two different theories of Malthus and Boserup. Make sure that you can clearly explain the two theories in detail, looking at both the positives and negatives in each case. Then finish your answer with a concluding statement stating which of the theories you think is stronger.

Knowledge check 8

Evaluate the theories of Boserup and Malthus.

The need for fertility policies (anti-natalist and pro-natalist)

The population of the world has been increasing at a fast rate over the last 100 years. Since 1960, the global population has gone from just over 3 billion people to 7.3 billion in 2015. This is an increase of 4.3 billion in 55 years. The UN estimates that the global population will continue to increase to over 9.5 billion by 2050.

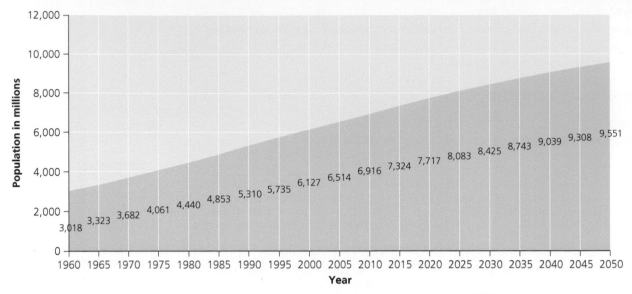

Figure 6 World population growth from 1960 and projected to 2050

This rapid rise in population has led to the United Nations Population Fund (UNFPA) supporting any measures that are aimed at reducing a country's birth rate.

Governments have tried to control their population levels in one of two ways.

Pro-natalist policies: these generally seek to increase both the population and the fertility rates in an area. In the past, places such as Nazi Germany used forceful governmental policy to create an 'Aryan race'. Tax allowances were given to anyone bearing more than one child. In the 1960s, Malaysian women were encouraged to 'go for five' in order to increase the number of potential workers, which would lead to a rise in development. In Singapore in the 1990s, the government was concerned that the total fertility rate (TFR) had fallen to 1.48. It therefore introduced new policies that would pay mothers more for maternity leave and pay parents cash gifts each year that they enrolled in the 'Baby Bonus Scheme'.

Anti-natalist policies: these have been more common in recent years as countries have tried to slow down their fertility rate. India started a programme from 1952 and other countries have followed suit, with many gaining the support of the UN. There is a wide range in the adoption of these policies — Mauritius has introduced a relaxed approach to its population crisis whereas other countries, such as China, have gone for a more bureaucratic approach.

Most countries recognise that there is a need to keep the population rising at a low, steady level in order to help the economy increase. This is often promoted by keeping the TFR above the replacement rate (see page 12). Governments might have to adopt particular indirect polices to 'encourage' people to continue to have babies when the TFR drops (Table 4).

Pro-natalist Policies aimed at increasing the population and fertility rates.

Anti-natalist Policies which aim to slow down the fertility rate.

Exam tip

It is really important that you have a firm understanding of the main differences between anti-natalist and pro-natalist fertility policies. Can you give more than one example of where these have operated?

Table 4 Policies that governments might use to alter population growth

Direct policies	Indirect policies	
Policy and laws	**Government spending**	
Minimum marriage age	Education	
Women's status	Primary healthcare	
Children's education and work	Family planning	
Breastfeeding policies	Incentives for fertility control/Maternity and paternity benefits	
Number of children per family	Old-age security	
	Tax programmes	
	Family allowances	
	User fees for larger families	

Case study

A fertility policy: China

China is a country that has gone through some staggering changes in its approach to fertility over the last 75 years (Figures 7 and 8).

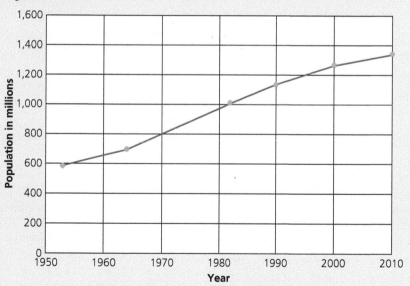

Figure 7 Population of China (from census figures)

Pro-natalist policy: the Great Leap Forward

In the 1950s and early 1960s China experienced a decline in population. This was mainly due to a number of natural disasters — earthquakes, hurricanes and floods which also led to a number of famines. President Mao Zedong introduced a programme to make China strong again, which would concentrate on transforming the country's economy from one based on agriculture to a more industrial approach. 'The Great Leap Forward', as the policy was known, reversed the downward trend in fertility and mortality, as 'a large population gives a strong nation'. Chinese population experts had suggested that the optimum population for China was around 700 million. The Cultural Revolution caused the population to increase at a rate of 55 million every 3 years (from 540 million in 1949 to 940 million in 1976).

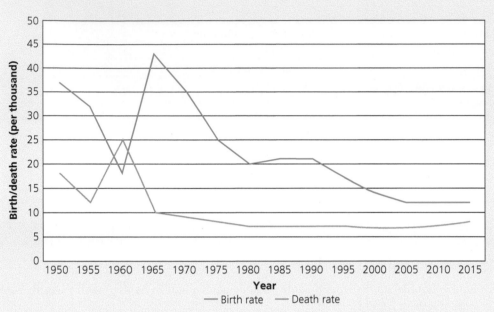

Figure 8 The demographic transition of China from 1950

Anti-natalist policy: the One-Child Policy (OCP)

In the 1960s in China, an estimated 42 million people died in famines and the Chinese government had to acknowledge that the increase in fertility was starting to create pressure on the balance of resources: China had to feed 22% of the world's population with less than 7% of the world's arable land. With the population quickly approaching 1 billion, despite the mass famines, a new, firmer approach to population control was required. State family planning programmes were introduced in the 1970s, such as 'Wan-xi-shao', which encouraged later marriages and longer gaps between children. The main aim of the OCP was to reduce the number of children being born per woman. The policy would also limit future demands for water and other resources, and raise standards of living.

The OCP was finally abandoned in January 2016 when the existing law was changed to a two-child policy.

The effects of the OCP

- Increase in female infanticide
  Males are favoured in Chinese society; boys can work in the fields and a daughter will look after the husband's family. A baby girl might be killed after birth to allow the parents an opportunity to have a boy in the future.
- Skewed male–female ratio
  In 2005, it was recorded that there were 118 boys to every 100 girls, showing an imbalance in the population. This will have social issues over time, as girls become more of a commodity.
- Ageing population
  Some demographers think that China will experience a severely ageing population from 2020 onwards. This will reduce the ratio of workers from 10:1 to 4:1 and put severe pressure on the continued economic improvements (the UN estimates that China will lose 67 million working-age people by 2030).
- 4-2-1 and 'Little Emperor' syndrome
  4 grandparents and 2 parents are focused on 1 child (4-2-1). This often means that the child is spoiled ('Little Emperor') and will not learn how to play and work with other children.
- Human rights
  Many of the methods used in the OCP affected a Chinese person's human rights. The Communist

regime in China was able to impose a control on its population in a way that few other countries would be able to replicate.

■ Demographic impact
Although the population of China now stands at over 1.3 billion people, it would be much larger if it were not for the OCP. One analyst suggests that from 1979 onwards 350 million abortions have taken place, which would have increased the population further. Population growth is, in fact, slowing down — the TFR was 2.98 in 1978 but only 1.54 in 2014.

■ Economic impact
The Chinese economy went through huge changes during the implementation of the OCP. In 1980, the GDP per capita was $310 and this has now increased to $7,924 in 2015. The Chinese people are much better off than they used to be. They have more disposable income and are looking to buy motor vehicles instead of bicycles, which is good for the economy but is putting new pressures on fuel supply and the environment.

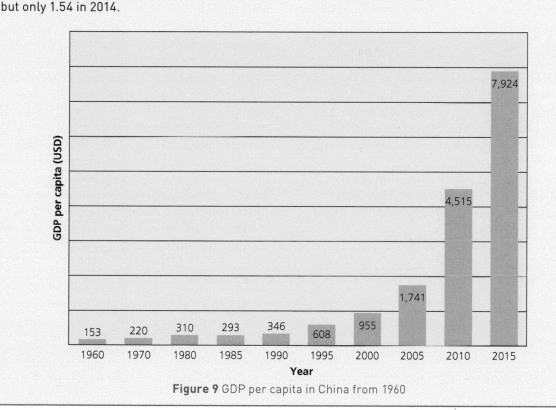

Figure 9 GDP per capita in China from 1960

Exam tip

Make sure that you understand the need for fertility policy in relation to the case study. How did this policy work? Was it successful in what it was trying to achieve? Facts and figures are vital for high marks in questions on this topic.

Describe some of the effects that the One-Child Policy had on China.

Summary

- There is a delicate balance between the population and the resources available in an area.
- The optimum population of an area is the number of people which, when working with all of the available resources, will return the highest standard of living.
- Overpopulation is when there are too many people relative to the resources and technology available to maintain an adequate standard of living.
- Underpopulation is when there are far more resources available in an area than the number of people living there.
- Carrying capacity is the largest population that the environment of an area can actually support with reference to the resources available.
- Sustainable development is 'development that meets the needs of the present without compromising the ability of future generations to meet their own needs'.
- A sustainable population is one that can be maintained at the number of people without affecting the quality of life or standards of living of that population.

- Thomas Malthus introduced a pessimistic theory, which argued that the demand for resources will be outstripped by population and this will become a crisis point in which a series of preventative/negative or positive checks will come into play.
- Ester Boserup wrote a more optimistic theory, which argued that as the population came close to the carrying capacity for resources, necessity would become the mother of invention and people would overcome the problem with technological innovation.
- Some countries have introduced pro-natalist policies in order to increase the size of the population while others have produced anti-natalist policies, which are aimed at decreasing the size of the population to fit the resources available.
- A case study of a country that followed an anti-natalist policy is China. In 1970 the One-Child Policy was introduced, which limited the number of children born to a couple. The policy was strictly followed in the Communist country but is widely thought to have had positive effects on the country and reduced the population overall, allowing the economy to grow.

Topic 2 Settlement and urbanisation

Settlement change

Distinguishing between rural and urban settlement

It is not always a straightforward process to work out the differences between rural and urban settlement, despite rural areas being very different to urban areas (Figure 10). As you move away from the central business district (CBD) in a city, the amount of rurality increases.

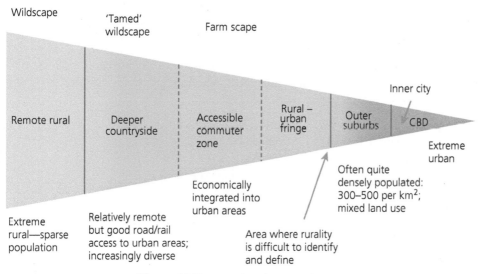

Figure 10 The rural–urban continuum

The extent to which an area is defined as rural or urban depends on a number of key factors:

- **Settlement size:** many countries have set criteria of how many people need to live in an area for a particular designation. In the UK, a settlement with a population of less than 1,000 is usually defined as being rural.
- **Population density:** an area with more than, say, 100 people per km² might be described as being urban, while one with less than 100 would be defined as rural.
- **Settlement function/land use/employment:** sometimes the function of an area will help to show whether it is more indicative of a rural or urban population.
- **Perception and service provision:** rural areas are characterised as having a dispersed population, with agricultural land use and fewer services available than in urban areas.

> **Exam tip**
>
> It can be difficult to describe the differences between rural and urban areas, so make sure that you have a clear understanding of how this works.

Table 5 shows an 'index of rurality', developed by Professor Paul Cloke, which identifies some indicators of rural life using available census data.

Table 5 Cloke's index of rurality

Indicator of rurality	Application to rural area
Population per hectare	Low
Percentage change in population	Decrease
Percentage of total population over 65	High
Percentage of total population male, 15–45 years old	Low
Percentage of total population female, 15–45 years old	Low
Occupancy rate: percentage of population at 1.5 people per room	Low
Households per dwelling	Low
Percentage in socio-economic groups 13/14/15, associated with farm work	High
Percentage resident for less than 5 years	Low
Distance from nearest urban centre	High

A settlement hierarchy can be used to rank the importance of places within an area. The population size of a place is the usual indicator used to determine how important a place might be and whether it sits within the rural or urban remit for settlement.

In terms of size, an isolated dwelling usually has a population of 2–4 people, a hamlet will usually have less than 100 people, a village less than 2,000 people and a small town less than 20,000 people. Larger settlements, like large towns, will have less than 100,000 people, cities between 100,000 and 1 million people and a conurbation will have more than 1 million people.

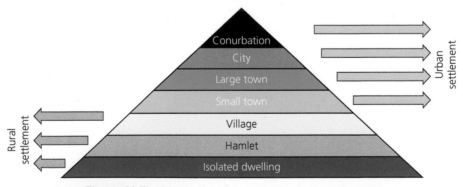

Figure 11 The hierarchy of rural and urban settlement

Often the most obvious difference between rural and urban areas is in the land usage. In Figure 12, Sue Warn summarises some of the factors that show the character of rural areas.

Knowledge check 10

What are the main factors that can influence the rural landscape?

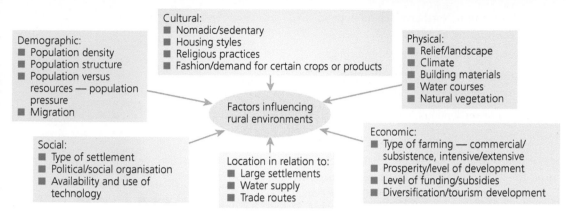

Figure 12 Factors that influence the rural landscape

Issues in the rural–urban fringe

The rural–urban fringe is the hinterland between a typically urban landscape and a typically rural landscape.

Greenfield developments

In the UK the growth of the 'suburbs' started in the 1920s. This is sometimes called 'urban sprawl'. In many towns and cities, new developments and housing estates are built at the edge of cities to reduce the pressure on the inner city. Land use planning around the edge of urban areas is critical to ensure that urban sprawl is contained and that land is used in the most effective way.

Greenfield development refers to an area of land surrounding a city or town that has not been developed or built up – but there are plans to change that. Green belt policies are often put in place to actively prevent urban sprawl into these areas. Planners use the concept of a green belt as an invisible line to prevent urban development into an area, seeing the green belt as the 'lungs' of an urban landscape.

In 1971, Belfast had a population of around 600,000 and over the next 20 years there was a slow but steady decline of the inner city population as people moved to the suburb areas to the north, east and south of the city. In 1964, there was an attempt to stop the continued sprawl of Belfast with the Matthew Regional Plan. Matthew noted the need for a 'greenscape', capable of sustaining agriculture, forestry and outdoor recreation, and that was convenient to the urban population.

Green belts have been quite successful in slowing urban sprawl. However, there have been a number of side effects.

- Green belts sometimes force development to take place further into the countryside.
- Inner-city areas can become more crowded.
- Competition for land increases within the city (and in areas surrounding the green belt), forcing land prices up.
- Longer commuting distances into city centres increase congestion and pollution, resulting in a need for improvements to the transport network.

Green belt an area of undeveloped land around a city on which building is restricted.

Knowledge check 11

Describe how a green belt might stop urban sprawl.

The Planning Strategy for Rural Northern Ireland notes that the strategic objectives for green belts in the country are as follows:

1 to prevent the unrestricted sprawl of large built-up areas,
2 to prevent neighbouring settlements from merging,
3 to safeguard the surrounding countryside,
4 to protect the setting of settlements,
5 to assist in urban regeneration.

Some developers are now beginning to argue that the 'brownfield' sites in urban areas that could be used for potential redevelopment are running out and that areas of greenfield will need to be earmarked for future construction.

Suburbanisation

Suburbanisation is often the first step in the decentralisation of the inner city. In 1964, Matthew commented that Belfast needed some of its slum areas cleared, and proposed measures to move people to the suburbs and beyond. Much of the subsequent suburbanisation began as extensions and improvements to the transport network and infrastructure (Figure 13).

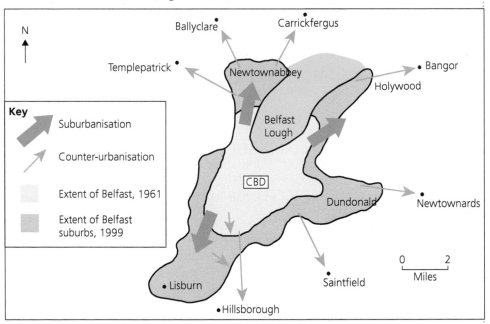

Figure 13 Suburbanisation and counter-urbanisation around Belfast

As a city grows outwards and encroaches on the land surrounding it, it changes the characteristics of the urban landscape. The developing suburban areas experience pressure as more people create more demand for services, housing, shops, parking and social facilities. Sometimes these areas expand to become merely 'dormitory towns' or 'commuter villages' from where people travel to work.

> **Exam tip**
>
> Make sure that you can explain both the positives and negatives of green belts.

> **Knowledge check 12**
>
> Describe and explain the difference between suburbanisation and counter-urbanisation.

Counter-urbanisation

If suburbanisation is the movement of people from the inner city to the suburbs, then counter-urbanisation is the movement of people from the inner city and suburbs to places beyond the city limits/metropolitan area. This usually involves movement to surrounding rural towns and villages that are within commuting distance of the city.

Since the early 1980s, people have continued to move out from the fringes of Belfast. These movements occurred as incomes allowed people to consider living in bigger, more expensive houses, with land and space around them.

Towns like Ballyclare, Templepatrick, Greenisland, Carrickfergus, Bangor, Newtownards, Comber, Moira and Hillsborough have all grown as a result of this counter-urban movement.

It would be difficult for any suburbanisation or counter-urbanisation to take place without good transport links to the places to which people are moving. Belfast already had a well-developed road and rail network, so people could choose to use their own independent transport or the public transport system.

What effects have these developments had?

The above movements have led to a number of key issues within the rural–urban fringe, including:
- traffic congestion and pollution,
- competition for land, raising house prices,
- the need to build and develop similar services to those in the city (e.g. shops, restaurants, entertainment) and
- continued sprawl into the countryside — removing green space and reducing habitats for wildlife.

Exam tip

Students often get the urban movements mixed up, so make sure that you read the question carefully before explaining the process.

Knowledge check 13

What impact does urban sprawl have on the rural–urban fringe?

Knowledge check 14

What potential effect might urban sprawl have on the city centre/inner city?

Summary

- The rurality of an area depends on settlement size, population density, settlement function/land use/employment, perception and service provision.
- The rural–urban continuum can be used to measure the extent of rurality from the CBD.
- It is getting increasingly difficult to tell where urban areas stop and rural areas start, this creates a rural–urban fringe area surrounding a city.
- Settlement hierarchy can be used to identify the different types of rural and urban settlement.

- A green belt can be used to protect the land around a city and will stop the city encroaching further. A greenfield development is an area of land surrounding a city or town that has not been developed or built up but there are plans to change that.
- Suburbanisation is the movement out of the inner city to the city suburbs. It is mostly caused by social and economic change within the city.
- Counter-urbanisation is when people move from the inner city and the suburbs to places out of the city limits/metropolitan area. This is facilitated by having a good transport network.

Planning in rural environments

Countryside planning and management

You need to be able to explain how planning is used to protect the countryside. There are a variety of government departments that have responsibility for making sure that the environment is protected.

In Northern Ireland, the Department of Agriculture, Environment and Rural Affairs has responsibility for all aspects of rural planning. In 2012 the Regional Development Strategy (RDS) 2035 was published. This is a long-term plan that aims to ensure that all areas in Northern Ireland benefit from economic growth and recognises the importance of key settlements as centres for prosperity and growth. Recent changes to legislative requirements have been made through the Planning Act (2011), which means that planning powers have been transferred to councils.

The Planning Strategy for Northern Ireland sets out a number of Regional Planning Policies. Policy Design Principles (DES1) look at Countryside Assessments, which are part of the preparation for development plans.

DES1 notes that:

> 'Northern Ireland is blessed with a rich diversity of countryside. The distinctive character is dependent on the combination of the many different elements of the natural and man-made landscape. A quality countryside is a very important resource and should be highly valued. It contributes significantly to the identity of rural Northern Ireland and is a source of enjoyment and inspiration. A high priority will be given to its conservation and enhancement.'

The main aim of these protection measures is to ensure that the countryside is conserved for the future. This is done by preserving or protecting an area and making sure that any natural resources are used wisely. This means that planners within a rural area need to consider carefully the ways in which the area can be used and need to make sure that any development is sustainable. Measures need to be taken in planning land use and recreation/tourism activities in order to make sure that the fabric of the landscape is not changed beyond all recognition.

Recreation is when people use their leisure time for enjoyment or pleasure and do not stay overnight, for example a day trip to go walking. These activities often attract visitors to rural areas and provide much-needed jobs and income for local people. For this reason, a trade-off between using the landscape as a resource and protecting the landscape needs to be found.

Tourists are drawn to an area because of its recreational resources. However, tourists may travel to an area and stay overnight for recreational, leisure or business purposes. This means that tourists require services and facilities that allow them to develop a 'home from home'.

There are a number of measures that can be used to manage the countryside: Areas of Outstanding Natural Beauty (AONBs), Areas of Special Scientific Interest (ASSIs) and National Parks. Within Northern Ireland there are eight AONBs, 47 national nature reserves, 43 special areas of conservation and 10 special protection areas.

Knowledge check 15

How many of each of the protected areas are there across Northern Ireland?

Areas of Outstanding Natural Beauty

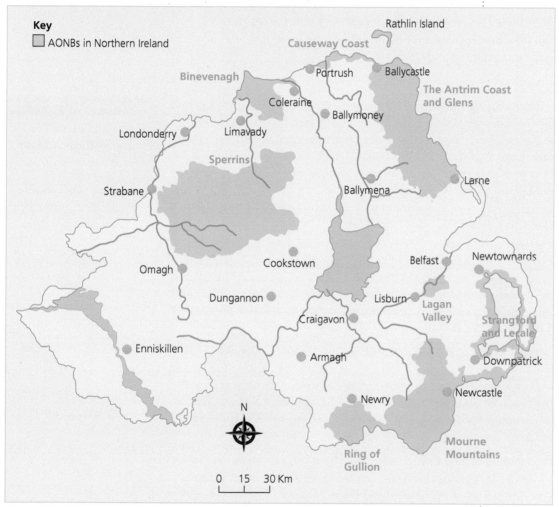

Figure 14 AONBs in Northern Ireland

An Area of Outstanding Natural Beauty (AONB) is described as being an area of the countryside that is considered to have landscape of significant value. It is a term used in England, Wales and Northern Ireland, and its main purpose is to conserve and enhance the natural beauty of a particular area. AONBs usually have a series of planning controls that help manage the area, while allowing people to enjoy it. There are 33 AONBs in England and Wales and eight in Northern Ireland (the most famous being the Antrim Coast and Glens, the Causeway Coast, the Mourne Mountains and Strangford Lough).

The Planning Strategy for Rural Northern Ireland deals with AONBs in Design Principle 4 (DES 4). The policy is mostly concerned with ensuring that any development proposals in AONBs are sensitive to the distinctive character of the area with a particular reference to the quality of the landscape, heritage and wildlife.

A total of 26% of Northern Ireland can be found within an AONB. These areas are more than places with amazing scenery — DES 4 lists that in Northern Ireland the main objectives are:

- to conserve or enhance the natural beauty or amenities of the area,
- to conserve wildlife, historic objects or natural phenomena within it,
- to promote its enjoyment by the public,
- to provide or maintain public access to it.

There is a clear understanding that the Department of Agriculture, Environment and Rural Affairs will not just protect (conserve) the unique qualities within an AONB, but also promote the enjoyment and use of the area (for recreation or tourism). The Department clearly notes that it will publish specific guidance on design principles to be applied in every AONB and that it has already done so for the Mournes and the Antrim Coast and Glens.

Knowledge check 16

Describe the main objectives for AONBs within Northern Ireland.

Areas of Special Scientific Interest

An Area of Special Scientific Interest (ASSI) is a conservation designation used in the UK in order to help protect some form of geologically or biologically significant area. The Northern Ireland Department of Agriculture, Environment and Rural Affairs describes ASSIs as 'protected areas that represent the best of our wildlife and geological sites that make a considerable contribution to the conservation of our most valuable natural places.'

The protection offered by designation as a 'Site of Special Scientific Interest' (SSSI) is usually for a small, localised area. It is often seen as a way of preventing the loss of biodiversity (plants and animals) and geodiversity (rocks, minerals, fossils, soils and landforms). In Northern Ireland, the Northern Ireland Environment Agency (NIEA) is in control of designating ASSI status, and it intended to have 440 ASSI areas declared by 2016.

The areas are protected using The Environment (Northern Ireland) Order 2002 Part IV (section 28). The Order declares that if such an area is designated an ASSI then the Department of Agriculture, Environment and Rural Affairs will make a statement about how the land will be managed (and include views on the conservation and enhancement of flora/fauna/features). Article 34 of the Order notes that land owners in an ASSI can enter a management agreement, which might impose some restrictions on what can and cannot take place in the area.

Some examples of ASSIs in Northern Ireland include The Gobbins in County Antrim, Carlingford Lough in County Armagh, The Copeland Islands in County Down, Cuilcagh Mountain in County Fermanagh, the River Roe and its tributaries in County Londonderry and Cashel Rock in County Tyrone.

National Parks

National Parks are managed carefully for conservation purposes. The term 'National Park' is an international designation covering around 6,555 parks worldwide. In the UK, National Parks are seen to have two central purposes: to conserve and enhance the natural and cultural heritage of the area and to promote understanding and enjoyment of the special qualities of the park by the public. There are 15 National Parks in the UK (10 in England, three in Wales and two in Scotland) and six in the Republic of Ireland.

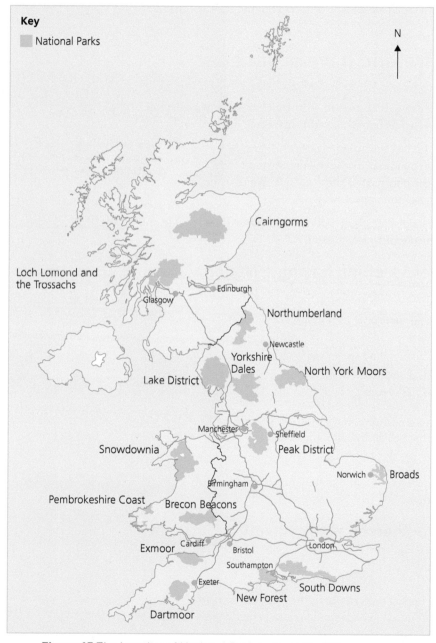

Figure 15 The location of National Parks across the UK

The aims of a National Park

In the UK, the 1995 Environment Act was a law that set out the two main aims of and purposes for National Parks in England and Wales:

1 to conserve and enhance the natural beauty, wildlife and cultural heritage,
2 to promote opportunities for the understanding and enjoyment of the special qualities of National Parks by the public.

When National Parks carry out these purposes they will also have a duty to:

■ seek to foster the economic and social well-being of local communities within the National Parks.

Exam tip

Make sure that you have a good knowledge of the different countryside management designations.

Should Northern Ireland have a National Park?

Rural areas are going through big changes at present — there is ongoing agricultural transition and tourism has been noted as being a key driver for sustainable rural development.

In 2012, the Northern Ireland Environment Minister, Alex Atwood from the SDLP, stated that: 'Northern Ireland is the only part of these islands that does not have National Parks and it is time to grasp the opportunity'.

He also went on to note that, if established, these parks would 'advertise the scale and wonder of our heritage and create jobs in times of need'.

Arguments for Northern Ireland having a National Park

Economic benefits
■ Potential £2–4 million additional funding
■ Direct employment of 30 people via a National Parks Authority
■ Increased opportunities for recreation and increased number of visitors
■ Increased visitor expenditure and employment associated with tourism, industry and countryside management

Social benefits
■ New regulations for housing and people living in these areas

Environmental benefits
■ Landscape and built heritage protection and maintenance of the areas of biodiversity

Arguments against Northern Ireland having a National Park

Economic issues
■ Any legislation might have an impact on farming in the area. Farmers fear that their practices would be compromised, which would restrict their opportunity for wealth generation.
■ Increase in number of second homes

- Increase in house prices and rates
- Change to employment profile — tourism jobs tend to be lower paid and seasonal

Social issues

- Conflict between tourism/recreation and landowners, especially if access points are not adequate
- Increases in traffic congestion associated with increasing numbers of visitors

Environmental issues

- Possible effects due to visitor numbers on the landscape, biodiversity and built heritage unless careful management is put in place

A Mournes National Park?

The Mournes area is noted by Ruth McAreavey (2010) as having,

> 'a fragile and fragmented economy and relies on agriculture, tourism, self-employment and commuting. Specific activities include farming, forestry, fishing, mineral extraction, water supply, tourism and recreation. Tourism related activities provide up to 15% of employment; meanwhile some 53% of the land is actively farmed and is in small-holdings.'

In September 2012 a series of public meetings were held in Newcastle, County Down to discuss the possibility of the area becoming the first designated National Park in Northern Ireland.

Jason Rankin from the Ulster Farmers' Union has said,

> 'Farmers are passionately opposed to these proposals. The farmers in England are restricted in so many ways which we can ill afford at a time when we're looking to the agri-food industry to take us out of recession in NI.'

Valerie Hanna of the Mournes AONB Residents' Action Group said,

> 'Large companies do not locate to National Parks. House prices rise and rates will rise because people look at National Parks as a status symbol where they can buy a second home.'

The Mourne Heritage Trust has welcomed any move to change legislation in Northern Ireland in favour of National Parks. It notes:

> 'As well as protection and enhancement of special qualities, the National Parks proposals — if designed and implemented effectively — offer a potential base for development of enhanced visitor and environmental management services and a stimulus to sustainable economic and social development ... The economic benefits are not all about tourism but can be derived by a broad range of economic sectors and enterprises.'

Knowledge check 17

Evaluate the positive and negative impact that a National Park might have in Northern Ireland.

Evaluating the arguments

A Northern Ireland National Park will help challenges to conservation

Soil and hillside erosion: hill walking, mountain biking, quadbiking and 4×4 off-road driving have all become popular in the area. Heavy use of paths can cause extensive damage to soil and hillsides. Overgrazing by sheep on hill areas can also reduce vegetation, leading to soil erosion.

Damage to wildlife: erosion as well as clearance of hedgerows to enable the extension of farmland or recreational facilities destroys wildlife habitats, including bird nesting sites. Litter can be a danger to animals.

Damage to farmland: walkers might leave gates open and livestock can be scared or attacked by dogs.

The management strategy

Protecting the landscape: a National Park Management Plan will protect the landscape by working with farmers and other land managers to encourage land management systems within the park. The character of the park would be enhanced. Heritage, historic buildings and archaeological remains are given statutory protection.

Protecting the biodiversity and ecosystems: the diversity of wildlife has been under threat due to the rapid rate of change, so the countryside will be managed more carefully. Healthy soils and watercourses are vital, so these would be maintained in high-risk areas. Local partnerships with protection organisations like the National Trust and the RSPB would be further enhanced.

A Northern Ireland National Park will help challenges to recreation

Congestion of villages and beauty spots: some of the most popular areas, known as 'honeypots', attract large numbers of visitors. This can lead to overcrowded car parks, blocked roads and overstretched local resources.

Effect of recreational activities: some activities cause damage and create conflict. For example, high-speed boats can cause excessive amounts of noise pollution and might conflict with fishing or swimming.

The management strategy

Sustainable approach to transport: cycle routes would be developed but occasional visitors may not use these to get into the park area. More sustainable local transport links into local towns and urban centres would be encouraged.

Management of recreational activities: many of the activities would be closely managed and controlled. Watersports would be limited, and fishing controlled through the involvement of fishing clubs. Motor sport activities (such as off-road/4×4 driving) would be limited to particular areas, while traffic-calming devices are being fitted to 'green lanes'.

A Northern Ireland National Park will help challenges to tourism

Effect on local services: services opened up to serve tourists, such as gift shops and cafes, can displace those serving the local population, including bakers and butchers. House prices rise as demand for second homes and holiday cottages increases.

Pressure on services: tourists require new hotels, holiday cottages and caravan parks. These take up valuable living space in the local area. Improved sewerage, electricity supplies, water supplies, phone lines and internet connectivity are all required to service the tourists.

The management strategy

Sustainable tourism: the Management Plan could aim to welcome people for 'escape, adventure, enjoyment and sustainability'. This involves management of the natural environment, heritage assets, local culture and local infrastructure so that the specific needs of visitors can be met.

Sustainable communities: the aim is for revenue from tourism to allow sustainable development within the local communities so that small settlements can adapt to new challenges, but still retain their historic and cultural identity.

> **Exam tip**
>
> Over the last 10 years there have been frequent discussions concerning whether or not Northern Ireland should have a National Park — no doubt these arguments will surface again. Make sure that you can demonstrate why a designation as National Park might be something that brings added protection and value to the area concerned.

Summary

- In Northern Ireland, the Department of Agriculture, Environment and Rural Affairs has responsibility for all aspects of rural planning.
- Rural environments need protection and various planning mechanisms are used to do this, including AONBs, ASSIs and National Parks.
- An Area of Outstanding Natural Beauty (AONB) is an area of the countryside that is considered to have landscape of significant value and is used to conserve and enhance the natural beauty of an area.
- An Area of Special Scientific Interest (ASSI) is a conservation designation used in the UK to help protect some geologically or biologically significant areas.

- National Parks are areas that are managed carefully for conservation purposes. The two main aims of and purposes for a National Park in the UK are as follows:
  1. to conserve and enhance the natural beauty, wildlife and cultural heritage,
  2. to promote opportunities for the understanding and enjoyment of the special qualities of National Parks by the public.
- In Northern Ireland, people are divided on whether a National Park would bring more positives than negatives.

Urban challenges

Issues and challenges of the inner city in MEDCs

You need to be able to demonstrate knowledge and understanding of **issues and challenges found in the inner city in MEDCs, including economic and social deprivation, re-urbanisation and gentrification,** with a case study to illustrate these urban issues (e.g. Belfast).

Belfast

Belfast grew rapidly in the 1800s with the introduction of textile and linen mills along the rivers of north and west Belfast (e.g. on the Shankill, Crumlin and York Roads). Engineering works (e.g. Mackies and Sirocco) grew up along the Springfield and Newtownards roads, and shipbuilding and its associated industries (like rope-making and furniture-making) developed near the mouth of the River Lagan (Figure 16).

Inner-city Belfast continued to grow rapidly. Much of the housing was built close to the factories in long, straight-terraced formations.

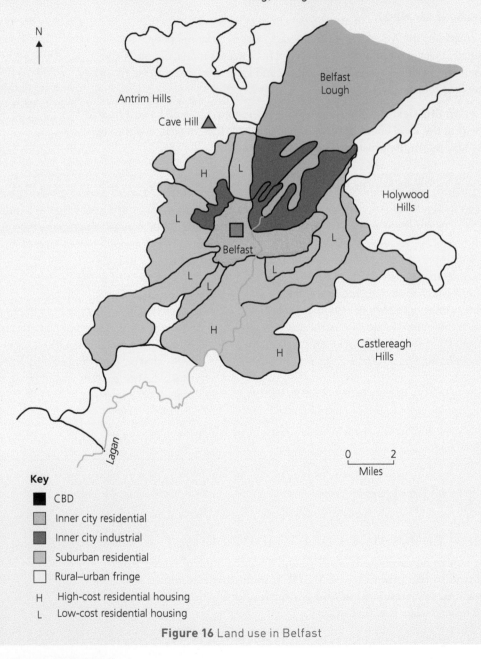

Key

- ⬛ CBD
- ⬜ Inner city residential
- ⬛ Inner city industrial
- ⬜ Suburban residential
- ⬜ Rural–urban fringe
- H High-cost residential housing
- L Low-cost residential housing

Figure 16 Land use in Belfast

Redevelopment is when an area is demolished and redesigned. In the inner city this might mean that a street of terraced houses is knocked down and replaced with a block of flats or other cheap housing.

Regeneration is when an area is upgraded. The aim is to improve the social and economic conditions. It happens in areas with issues of dereliction, pollution and out-migration. It might involve restoring old buildings and redesigning them for a different use.

Gentrification occurs when an area is demolished and upgraded, attracting richer people to live in the new, modern, expensive properties.

Economic and social deprivation

In the 1960s and 1970s, houses within inner city Belfast were in poor condition. Much of the heavy industry that had employed people was in decline. Factories were closing and work was increasingly scarce. Employment became less centralised and workers sometimes had to travel to the edge of the city. Many had to move out, while those who were left had less money to spend, and shops and services began to close down.

Deprivation is defined as 'unmet need across a number of domains'. It is not limited to poverty, a lack of money or material possessions.

According to the most recent Northern Ireland *Multiple Deprivation Report* (produced by NISRA, 2010), some of the areas of highest relative deprivation in Northern Ireland were found in inner city areas — the Falls, New Lodge, Shankill, Crumlin Road, Ardoyne, Upp er Springfield and Whiterock.

This report was generated using 52 separate indicators of deprivation, which are divided up into: income (25%), employment (25%), health and disability (15%), education, skills and training (15%), proximity to services (10%), living environment (5%), and crime and disorder (5%).

Some of the most up-to-date deprivation figures illustrate the issues in the inner city.

Economic data

Unemployment: the current unemployment rate for Northern Ireland (those of working age who are claiming unemployment) is 5.9%. The rate for Belfast is 7.4% (13,816 people). In north Belfast it reaches 16.9% in Duncairn and 13.5% in New Lodge, while in west Belfast's Shankill it is at 13.5%.

Further aspects of deprivation can be noted from the amount of incapacity benefit, income support, housing benefit, pension credits and disability allowances paid to residents.

Social data

Free school meals: one useful social measure is the extent of free school meals in an area. This is a guide to the number of families who are on low incomes. In 2012/13, 40% (5,333) of students attending secondary (non-grammar) schools received free school meals. This compared with 7.4% of grammar school students.

School attendance: another measure used is absenteeism (% of total half days) at secondary schools. In 2013 the Falls recorded 8.8%; Duncairn and Shankill were both at 11.2%.

Qualifications: a final measure of deprivation is the proportion of working-age adults with no qualifications. For example, in north Belfast's Water Works 32% of the population have no qualifications; in Duncairn the rate is 41%.

The people in these areas are in a cycle of poverty and struggling to break out of it. Disadvantaged children are more likely to fail at school, will leave with few qualifications and will be less likely to get a good job.

Re-urbanisation

Re-urbanisation is the movement of people back into an area that had previously been abandoned. The key feature of any re-urbanisation is that people from the suburbs or from outside the city will make a decision to move back into the city. Usually the stimulus for any movement will come as a result of government intervention to improve an area.

The development of the Titanic Quarter and Laganside (also an example of gentrification) are both attempts to re-urbanise areas of Belfast that experienced serious decline.

Titanic Quarter

Much of the traditional manufacturing base in twentieth-century Belfast was located in Queen's Island. The decline of Harland and Wolff (a British heavy industrial company) from over 20,000 workers to 500 by 2002 meant that there was a large area of derelict land in the Belfast inner city (around 185 acres), with little opportunity for employment or economic growth.

Building started in 2006 with the aim of creating a new, fresh and modern space where people would be able to come to live, work and play. The hope was that new transport and communication links would allow this area to be the most accessible in the city, and that this development would breathe new life into Belfast. Key aims included:
- the building of up to 5,000 dwellings
- the creation of a high-quality business area
- commercial development
- a Titanic signature project
- a major educational, third-level campus (Belfast Metropolitan College)
- restoration and conservation of the former Harland and Wolf headquarters
- the creation of hotels and other tourist accommodation
- the development of new leisure facilities, restaurants, cafes/bars and health clubs.

In many ways, the timing of the first release of residential accommodation took place at the wrong time because the global financial crisis and recession took hold and much of the proposed development has had to be scaled down in recent years. However, the expansion of the area continues, with renewed plans for more accommodation to be built in the near future.

Gentrification

Gentrification (Table 6) is the development of run-down areas of a city so that they become fashionable places to live. Often the redevelopment means that the people who originally lived in these urban areas cannot afford to buy the new properties and are forced to move to other parts of the city.

Laganside

The Laganside development started at the Lagan Weir, which is used to control the amount of water in the river upstream. The Laganside Corporation became responsible for developing 140 hectares of land alongside the River Lagan and 70 hectares in the Cathedral Quarter.

By the time the project was completed the Corporation estimated that over 14,200 permanent jobs had been created. A total investment of £939 million resulted in the creation of 213,000 m² of office space, 83,000 m² of retail and leisure space and 741 housing units.

Some houses in the Mays Meadow area and at Lands End Street, Laganview Street and Newfoundland Street (all around Bridge End) were removed and new apartment blocks were put in their place.

Many of the communities who lived beside the Laganside development felt that they had been ignored and felt no positive impact from the regeneration.

Table 6 Features of gentrification

Positive features	Negative features
■ Houses will be improved — old, dilapidated buildings will be regenerated. ■ House values will increase. ■ New businesses to service the new, richer community will be set up. ■ Crime rates can fall.	■ A high demand for houses can cause problems. ■ Increased house prices mean that original residents are stuck in the area or move out, and their children cannot afford to live in the same area. ■ There can be conflict between the original and new residents.

Exam tip

For both of the urban case studies you need to make sure that you can demonstrate a thorough knowledge and understanding of the place you are talking about. Use place names, road names and facts and figures to show your command of the answer.

Redevelopment When an area is demolished and redesigned.

Regeneration When an area is upgraded or renewed.

Gentrification When an area is redeveloped and upgraded, attracting richer people and after displacing poorer families.

Exam tip

Make sure that you understand the difference between re-urbanisation and gentrification, and be ready to give examples of different projects around Belfast to support your answer.

Knowledge check 18

What is the difference between redevelopment and gentrification?

Exam tip

It is essential that you know facts and figures to show an understanding of the level of poverty in the inner city.

Issues and challenges of the inner city in LEDCs

You need to be able to demonstrate knowledge and understanding of the **issues and challenges found in LEDC cities, including the growth of informal settlements, service provision and economic activity**, with a case study to illustrate these issues (e.g. Nairobi, Kenya).

In the 1970s and 1980s many LEDC cities experienced a rapid expansion as many poor people left their small patches of land in the countryside and migrated into the urban areas. This, in turn, created huge problems for the cities, which had to grow to cope with the influx of people.

Generally, when an LEDC city expands like this, an economically segregated area develops on one side of the CBD, with expensive accommodation including high-rise apartments and some gated communities. Adjacent to this, and found on more marginal land, poor-quality housing develops into shanty towns or slums.

Wedges or ribbons of development encourage factories to be built — usually along the routes of major roads (towards ports or airports) or along railway lines. A surrounding ring of more mature suburbs/medium-cost housing is often found beyond this (Figure 17).

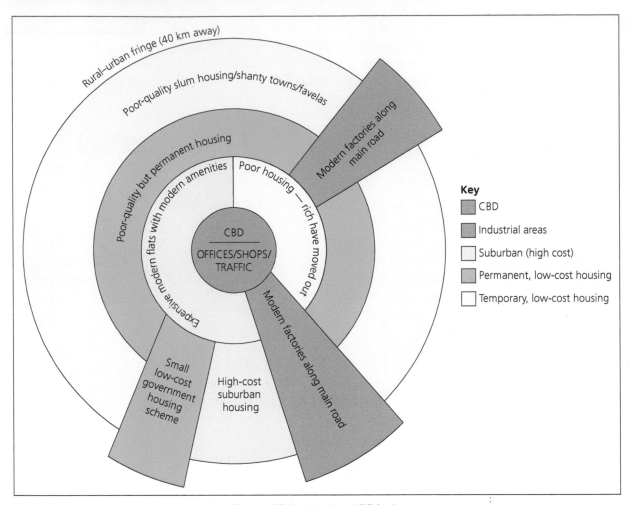

Figure 17 Model of an LEDC city

Unlike in MEDC cities, parcels of land on the outskirts (rural–urban fringe) are used to house the most affluent in society, but these are often interspersed with slum areas that develop near rubbish tips and poor-quality land.

The development of the LEDC cities in this way has led to a number of major problems.

Case study

Nairobi, Kenya

Many African cities were originally built as colonial staging posts. The British colonial authorities set up Nairobi in 1899 as a railway camp on the Mombasa to Kisumu line. However, the city developed quickly and by 1907, it was established as the capital of Kenya. Nairobi's population grew to 1.3 million in 1990, 2.1 million in 2000 and by 2011, was estimated at 3.3 million.

The growth of informal settlements

One of the most common features of any large urban settlement in an LEDC is the growth of informal settlements — 'shanty towns' or slums. These are usually found in low-lying areas where there is a risk of flooding, with poor soils, close to rubbish tips, or on steep slopes where landslides might be common.

Few people in slums own the shacks they live in. The owners are usually the original builders of the dwellings, which means that they are poorly constructed with whatever materials can be gained in the local area. Typically houses are small — about 3m × 3m — with a floor made of mud. There might be a small window on more advanced properties but usually the construction of mud and wattle or corrugated tin will not allow for this.

Kibera slum

One infamous slum in Nairobi is Kibera. It has an estimated population of 170,000 people (though some sources estimate the population to be as high as 1 million) who are extremely poor. An estimated 2.5 million of the Nairobi population live in some sort of slum area, representing over 75% of the total population.

Few people have jobs and those who do earn little money. There are many social issues, such as a high level of HIV infection and regular cases of assault and rape. Education is sparse because people have little money to pay for schooling. Clean water and good sanitation are non-existent and life expectancy is low because people live in these squalid conditions with no access to decent healthcare.

Service provision

The lack of jobs and space for people to live means that there is little money to invest in building a decent infrastructure and providing access to services. Water supplies are inconsistent and often require people to collect drinking water from the same places that are used for sewage disposal. Waste disposal services are entirely lacking, which helps to spread disease.

The Kenyan government owns the land in Kibera, but it refuses to officially acknowledge the settlement, which means that the development of basic services like schools, clinics and toilets is limited. Shops are basic, with any food sold being old, poorly stored and exposed to heat and contamination.

Until recently there was little access to water or electricity in Kibera. UN-Habitat has been involved in getting electricity to about 20% of the residents, but it can be expensive. Any medical care is provided by charities and religious groups — global charities have set up clinics where people can have a free HIV test and gain access to free anti-retroviral (ARV) treatments if they are found to be HIV positive.

Economic activity

Jobs are hard to come by in the slum. Most of the people lack any form of education and are unskilled. There are few 'formal' jobs for people so the ever-increasing population has to resort to 'informal' service-related jobs in order to survive. These include washing car windows, picking out recycling materials at the city dumps and prostitution, all of which are usually carried out at great personal risk and often with very little reward.

Some transnational companies set up factories in LEDC cities to benefit from the cheap labour. The available workforce is so big that employers can demand difficult and binding conditions, with employees happy to work hard for little reward just so they can stay off the streets.

Summary

- In MEDC cities, the inner-city zones often present unique challenges in relation to economic and social deprivation.
- Poverty can have a big influence on lifestyles, education, job prospects and life expectancy.
- Inner-city areas require government intervention to get the area out of a 'spiral of decline', so re-urbanisation and gentrification projects are used (such as the Titanic Quarter and Laganside in Belfast).

- In LEDC cities, the challenges of urban living are caused by rapid urbanisation (and in-migration).
- This rapid urbanisation creates pressure on space, which leads to a growth of informal settlements/slums throughout the city and problems of service provision.
- The rapid urbanisation can create both opportunities and issues for the economy (e.g. Kibera in Nairobi, Kenya).

■ Topic 3 Development

Measuring development

The problems with defining 'development'

There is no agreed definition for 'development'. When we look at the world we live in, we might use certain standards to indicate the advancement of a particular place. Most geographers accept that the study of development considers the quality of life for people living in a particular area. This will take into consideration wealth, aid, healthcare, education, poverty, infrastructure, political landscape, economics and environment.

The World Bank notes 298 separate indicators that can be used to analyse the differences between countries. This helps to highlight the division there is across the world. Over the last 200 years a 'development gap' has opened up between the rich MEDCs and the poorer LEDCs.

Economic measures of development

Economic measures of development help us to work out how much money or wealth there is within a country and how people actually earn that wealth. This is one of the easiest ways of measuring how people from one country compare with people from other countries.

Gross national product (per capita) or gross national income

Gross national product (GNP) per capita, or gross national income (GNI), measures the total economic value of all of the goods and services provided in a country through a year, divided by the number of people who live in that country. The amount is worked out in US dollars so that an easy comparison can be made. The higher the GNP, the more developed the country. At present the country with the highest GNI (with $87,030 per person) is Qatar, while the lowest GNI is the Democratic Republic of the Congo, with a GNI of $350 per person.

Positive issues

- The comparison between countries is easy to understand.
- It gives a baseline of the amount of money in dollars per head of population that is earned within the country.
- It uses a common currency as a benchmark, which simplifies global comparisons.

Negative issues

- Working out how much money is earned within a country in a year is increasingly complicated to do — especially with the global markets.
- Changes in currency rates can affect it — 1 dollar in one country might not buy the same amount of food as 1 dollar in another country — so the real purchasing power of money can vary from one place to another.

Exam tip

Make sure that you have a good understanding of what development is but also note that coming up with one comprehensive definition is difficult.

Exam tip

GNP per capita or GNI is often the most quoted measure of development in the exam, so make sure that you can describe and evaluate it fully.

Knowledge check 20

What is the difference between GNP per capita and GDP?

■ It is a crude method of working out how much money is in a country's economy, and it fails to highlight that the distribution of this money within the country is uneven.

Percentage of people employed in primary activities

Sometimes we can learn how developed a country is by looking at the percentage of the workforce employed in the primary sector. A rich, more developed country is likely to have more people working in the secondary, tertiary and quaternary sectors. For example, the UK has around 2% of the population working in agriculture, while Vietnam has 73%.

Social measures of development

Social measures of development are used to assess how well a country is developing in areas that affect people directly, for example healthcare, education and diet. These measures help to indicate the quality of life of individual people in a country.

Healthcare

Life expectancy is a good example of a measure of healthcare. It can be affected by wars, disease and natural disasters, but can show how developed the medical facilities in a country are. The general rule is that the higher the life expectancy, the more developed a country is. People in the UK have a life expectancy of 81 years, whereas life expectancy in Chad is 48 years.

Infant mortality rate measures the number of children who die before they reach the age of 1 (from every 1,000 live births, per year). This helps to determine the quality of ante- and post-natal services in a country. Generally, less developed countries spend less money in tackling issues affecting infant mortality, while more developed countries spend more and have a much lower rate. In the UK this figure is around 6 per 1,000 live births per year, whereas in Sierra Leone the rate is 195.

Positive issues

■ Life expectancy and infant mortality rates are easy to work out. They depend on good vital registration data, which can be found in most countries.
■ They both give a useful figure that allows direct comparison with other countries and inferences to be made about the available health services.

Negative issues

■ Not all countries have robust methods of recording these vital registration data and in very poor countries information might be inaccurate.
■ Some aid organisations argue that variation in the amount of money allocated to tackling particular health issues might not give a true picture of the overall state of healthcare in a country. For example, political decisions might have been taken to not spend money on ante-natal care.

Knowledge check 21

Name one other potential economic measure that could be used to indicate development.

Life expectancy The average lifespan that someone born in a country can expect.

Infant mortality rate The number of children who die under 1 year of age per 1,000 live births per year.

Exam tip

There are many different social issues that can be used as measures of development — healthcare is an important one to evaluate, but make sure that you know at least one more social measure in detail.

Knowledge check 22

How does life expectancy help us understand healthcare as a social measure of development?

Education

One social measure of education is the adult literacy rate. This is the percentage of the adult population that is able to read and write. In the UK, and most other MEDCs, over 99% of adults have learned to read and write, yet in Somalia only around 24% of adults have achieved this.

Diet

Increasingly, international agencies such as the United Nations (UN) and the World Health Organization (WHO) have been using calorie intake to measure variation in diet across the world. In wealthy countries like the USA the average calorie intake per person is 3,725 kilocalories per day, which contrasts with poorer nations like Eritrea (1,555 kilocalories).

Composite measures of development

The human development index

In response to an over-reliance on simple measures of development, the UN looked for a measure that would combine some of the major indicators into one easy-to-use measure. The human development index (HDI) is worked out using three measures: life expectancy (a social measure), education (the mean number of years of schooling — a social measure) and gross national income per capita (an economic measure).

Each of the three measures that make up the HDI is ranked so that a high performance will be given an index mark approaching 1 and a poor performance will be towards 0. Each year the UN Development Programme (UNDP) updates its *Human Development Report*, which ranks countries across the world (Tables 7 and 8).

> **Exam tip**
>
> The HDI composite measure is a popular one in exam papers so make sure that you are well practised in being able to describe and evaluate it. However, questions might also test your knowledge of a second composite measure, so be ready!

Table 7 Top five countries (very high HDI, 2011)

Country in rank order	HDI value	Life expectancy (years)	Mean years of schooling (years)	Expected years of schooling (years)	GNI per capita ($PPP)
1 Norway	0.943	81.1	12.6	17.3	47,557
2 Australia	0.929	81.9	12.0	18.0	34,431
3 Netherlands	0.910	80.7	11.6	16.8	36,402
4 USA	0.910	78.5	12.4	16.0	43,017
5 New Zealand	0.908	80.7	12.5	18.0	23,737

Table 8 Bottom five countries (very low HDI, 2011)

Country in rank order	HDI value	Life expectancy (years)	Mean years of schooling (years)	Expected years of schooling (years)	GNI per capita ($PPP)
183 Chad	0.328	49.6	1.5	7.2	1,105
184 Mozambique	0.322	50.2	1.2	9.2	898
185 Burundi	0.316	50.4	2.7	10.5	368
186 Niger	0.295	54.7	1.4	4.9	641
187 Congo (DR)	0.286	48.4	3.5	8.2	280

GNI per capita ($PPP) is the GNI per capita based on purchasing power parity (PPP). This converts the usual GNI into international dollars so that an international dollar has the same purchasing power as a US dollar in the USA.

Positive issues

■ HDI takes into account both social and economic data and the social data include figures based on health and education, making this a more rounded and realistic measure compared with merely social or economic indicators.

■ The information is updated annually, which allows countries to monitor their development progress and smooth out any fluctuations from one year to the next, which means that the HDI position of countries will be unlikely to change much from one year to the next.

Negative issues

■ Some argue that wealth has too much importance within the HDI and that this can adversely influence the ranking of a country within the table.

■ Some suggest that the HDI takes too simplified an approach to measuring development and note that other factors beyond education, life expectancy and GNI are needed to get a more rounded picture of what it is like to live in a particular country (for example, by referring to markets, industry or environmental issues).

The physical quality of life index

This is another composite measure that uses life expectancy at year 1, infant mortality rate and basic literacy rate — each having an equal weighting. The values range from 0 to 100 and countries are ranked accordingly.

The physical quality of life index (PQLI) was originally developed as an alternative to GNP, but the UN HDI became more widely used. PQLI was often criticised because the three strands of the measure are similar and depend on each other: a low life expectancy indicates that money is not spent on healthcare, which will cause a high infant mortality rate. A low literacy rate indicates that money is also not spent on education of young people.

Knowledge check 23

What are the three main indicators used in the HDI?

Knowledge check 24

Why is the HDI a better measure than the PQLI?

Summary

■ The concept of development is not easily defined but it usually relates to the quality of life that people experience in an area.

■ Two key economic measures of development are gross national product (GNP) (per capita) or gross national income (GNI) and the percentage of people employed in primary activities.

■ Two key social measures of development are healthcare (including life expectancy and infant mortality rate) and education (literacy rate).

■ Two key composite measures of development are the human development index (HDI) and the physical quality of life index (PQLI).

Reducing the development gap

The Millennium Development Goals

In the year 2000, leaders of 189 nations agreed a new landmark commitment to 'spare no effort to free our fellow men, women and children from the abject and dehumanising conditions of extreme poverty'. The Millennium Development Goals (MDGs) were established, which would provide a framework for eight goals (and 23 specific targets) aimed at improving the lives and future prospects of more than 1 billion people by the year 2015.

The eight MDGs are:

1 eradicate extreme poverty and hunger

2 achieve universal primary education

3 promote gender equality and empower women

4 reduce child mortality

5 improve maternal health

6 combat HIV/AIDS, malaria and other diseases

7 ensure environmental sustainability

8 develop a global partnership for development.

Knowledge check 25

State the eight Millennium Development Goals (MDGs).

Now that we have reached the end of the MDG period, many within the UN believe that the world community has reason to celebrate. The MDGs have been seen to have saved the lives of millions and improved the standards of living for many more.

What effect have the MDGs had on improving global development up to 2015?

Goal 1: eradicate extreme poverty and hunger

- Extreme poverty has been significantly reduced over the last 20 years. In 1990 over half of the population in LEDCs lived on less that $1.25 a day and this dropped to 14% in 2015. This was still around 300 million workers living below the poverty line in 2015.

- The number of people living in extreme poverty declined from 1.9 billion in 1990 to 836 million in 2015. This means that more than 1 billion people have been lifted out of extreme poverty since 1990. However, over 800 million people are still living in extreme poverty.

- The number of people living on more than $4 a day tripled to 50% of the workforce in 2015 (up from 18% in 1991).

- The amount of undernourished people in LEDCs fell from 23.3% in 1990 to 12.9% in 2015. However, about 795 million people are estimated to remain undernourished (including more than 90 million children under the age of five).

Goal 2: achieve universal primary education

- The primary school enrolment rate for LEDCs reached 91% in 2015, up from 83% in 2000. The number of children not in school fell from 100 million in 2000 to 57 million in 2015.

- Areas of Sub-Saharan Africa noted the biggest improvement in primary education — their net enrolments rose by 20% between 2000 and 2015.

- Literacy rates among young adults aged 15–24 increased globally, from 83% in 1990 to 91% on 2015.

- However, in LEDCs, children in the poorest households are still four times more likely to be out of school than those in the richer households.

Goal 3: promote gender equality and empower women

- There are more girls in education now compared with 15 years ago. In South Asia in 1990, only 74 girls were enrolled in primary school for every 100 boys. This rose to 103 girls to every 100 boys in 2015.

- Women now make up 41% of paid workers outside of agriculture (from 35% in 1990).

- The proportion of women in vulnerable employment between 1990 and 2015 declined by 13%.

- Women still have to come to terms with significant gaps in terms of poverty, labour market and wages, as well as participation in decision making in the public sector.

Goal 4: reduce child mortality

- Across the world mortality rates for children aged under five declined by more than half — from 90 to 43 deaths per 1,000 live births from 1990 to 2015. The number of corresponding deaths declined from 12.7 million to 6 million.
- Measles vaccinations helped prevent 15.6 million deaths from 2000 to 2013 and the number of measles cases across the world declined by 67%.
- Children aged under five in rural areas are about 1.7 times more likely to die than those in urban areas.
- Children of mothers who have experienced secondary or higher education are three times more likely to survive than children of mothers with no education.
- Every day throughout 2015, 16,000 children under five continued to die from preventable illnesses.

Goal 5: improve maternal health

- Since 1990, maternal mortality ratios have declined by 45% worldwide. In South Asia, the ratio declined by 64% between 1990 and 2013 while in Sub-Saharan Africa it fell by 49%.
- 71% of births were assisted by skilled health professionals in 2014, an increase from 59% in 1990.
- The use and availability of contraception among women aged 15–49 increased from 55% in 1990 to 64% in 2015.
- Globally, there were an estimated 289,000 maternal deaths in 2013.
- There has been slow progress in ensuring that pregnant women get the right amount of ante-natal care.

Goal 6: combat HIV/AIDS, malaria and other diseases

- The number of new HIV infections fell by 40% between 2000 and 2013 (from 3.5 million cases to 2.1 million cases).
- By June 2014 there were 13.6 million HIV patients who were receiving anti-retroviral therapy (ART). This is a huge improvement on the 800,000 people in 2003. These treatments averted 7.6 million AIDS-related deaths between 1995 and 2013. In 2013 alone, the number of people getting ART rose by nearly 2 million in LEDCs.
- There were still an estimated 35 million people around the world suffering with HIV in 2013.
- 75% of new HIV infections in 2013 took place in 15 countries.
- A staggering 900 million insecticide-treated mosquito nets were delivered to malaria-endemic countries in Africa between 2004 and 2014. This led to a 58% decline in malaria mortality rates globally from 2000 to 2015. A total of 6.2 million deaths from malaria have been averted.
- New advances in tuberculosis diagnosis and treatment saved an estimated 37 million lives between 2000 and 2013. Tuberculosis mortality has fallen by over 45%.

Goal 7: ensure environmental sustainability

- Substances that could have caused ozone depletion have been virtually eliminated since 1990 and the ozone layer is expected to recover by 2050.
- The number of protected areas has increased since 1990.

- Forests are a safety net, especially for the poor, but they continue to disappear at a fast rate. However, new forest loss was reduced from 8.3 million hectares annually in the 1990s to 5.2 million hectares in 2010.

- Between 1990 and 2012, the amount of global carbon dioxide emissions increased by over 50%.

- 91% of the global population had access to clean drinking water in 2015, compared with 76% in 1990. A total of 1.9 billion people have gained access to drinking water in their own homes, meaning that 58% of the global population have this service.

- 2.1 billion people now have improved sanitation and the proportion of people practising open defecation has fallen by half since 1990. However, 2.4 billion people still have to use poor sanitation facilities and 946 million people still practise open defecation.

- The proportion of the urban residents living in slums in LEDCs fell from 39.4% in 2000 to 29.7% in 2014. However, more than 880 million people are estimated to be living in slums today (compared with 689 million in 1990).

Goal 8: develop a global partnership for development

- Official development assistance (aid) from MEDCs improved by 66% between 2000 and 2014, reaching $135 billion (see Figure 18).

- Bilateral aid to LEDCs fell 16% by 2014 (to $25 billion).

- In 2014, countries like the UK, Sweden, Norway, Denmark and Luxembourg exceeded the UN official assistance target of 0.7% of GNI.

- In 2014, 79% of imports from LEDCs to MEDCs were admitted duty free (up from 65% in 2000).

- The amount of money payable to service international debts fell from 12% in 2000 to 3% in 2013.

- By 2015, a mobile/cellular signal covered 95% of the world's surface.

- Internet provision grew from 6% of the global population in 2000 to 43% in 2015.

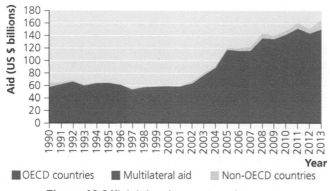

Figure 18 Official development assistance to LEDCs, 1990–2013

There is little doubt that the MDGs have had some impact on the world — however, some of the most poor and vulnerable people continue to be left behind, especially in relation to the following issues.

- Gender inequality persists.

- Big gaps exist between the poorest and the richest households, and between rural and urban areas.

- Climate change and environmental degradation undermine progress made, which causes poor people to suffer more.
- Conflicts remain the biggest threat to human development.
- Millions of poor people still live in poverty and hunger, without access to basic services.

A final assessment of how much impact the MDGs have had is very complicated due to the number of targets and countries involved. Table 9 shows the achievements for two regions and three targets. East Asian countries (EAPs) have achieved the extreme poverty target, while 18 countries in Sub-Saharan Africa (SSA) are nowhere close to their target. Yet, both areas have made less progress in relation to hunger or maternal mortality.

Table 9 Progress towards the MDGs in two regions

Progress (number of countries in region)	Extreme poverty		Hunger		Maternal mortality	
	SSA	EAP	SSA	EAP	SSA	EAP
Target met	11	9	11	5	0	1
Sufficient progress	5	1	4	2	4	4
Insufficient progress	5	0	4	1	4	1
Moderately off target	2	1	1	0	16	3
Seriously off target	18	1	24	8	21	10

The 2030 Agenda for Sustainable Development

On 25 September 2015, the UN adopted a set of goals that were aimed at ending poverty, protecting the planet and ensuring a new prosperity for all. This was part of a new 15-year Agenda for Sustainable Development. The UN General Assembly noted that it was:

> 'resolved to free the human race from the tyranny of poverty and want to heal and secure our planet. We are determined to take the bold and transformative steps which are urgently needed to shift the world on to a new sustainable and resilient path.'

The 17 Sustainable Development Goals (SDGs, also called the 'Global Goals', see Figure 19) and 169 targets have been built on the success of the Millennium Development Goals (MDGs) and aim to complete what the MDGs did not achieve and end all forms of poverty. The new goals are unique in that they call for action from all countries while also protecting the planet.

There is a much wider scope for the SDGs compared with the MDGs — their aim is to address the root causes of poverty, and the need for development that works for all people. The SDGs are also much more ambitious than the MDGs; they aim to address inequalities, economic growth, decent jobs, climate change, sustainable consumption, peace and justice, and issues affecting cities and human settlements, industry, oceans, ecosystems and energy.

A major difference is that the SDGs are to be applied universally to all countries, whereas the MDGs were intended for action in LEDCs only.

Exam tip

Make sure that you develop a good understanding of the positive and negative effects that have come about as a result of the MDGs. You might need to take a look at what the targets set underneath each goal were.

Knowledge check 26

How effective have the MDGs been since their implementation in 2000?

The SDGs also have a strong focus on how the goals will be implemented — there is a wider discussion about the mobilisation of financial resources, capacity-building and technology.

The UN notes that the Global Goals are aimed at realising human rights for all, achieving gender equality and empowerment within a sustainable development context focused on economic, social and environmental factors. They are designed to stimulate action over the next 15 years for:

- **People:** the UN is determined to end poverty and hunger in all its forms, and to ensure that all humans can fulfil their potential in dignity and equality, in a healthy environment.
- **Planet:** the UN is determined to protect the planet from degradation, including through sustainable consumption and production, managing its natural resources and taking urgent action on climate change, so that it can support the needs of present and future generations.
- **Prosperity:** the UN is determined to ensure that all humans can lead prosperous and fulfilling lives and that economic, social and technological progress occurs in harmony with nature.
- **Peace:** the UN is determined to foster peaceful, just and inclusive societies that are free from fear and violence. There can be no sustainable development without peace and no peace without sustainable development.
- **Partnership:** the UN is determined to mobilise the means required to implement the agenda through a renewed Global Partnership for Sustainable Development.

Figure 19 The Sustainable Development Goals ('Global Goals')
Source: United Nations

Table 10 The Sustainable Development Goals ('Global Goals')

Goal 1	End poverty in all its forms everywhere	Goal 10	Reduce inequality within and among countries
Goal 2	End hunger, achieve food security and improved nutrition, and promote sustainable agriculture	Goal 11	Make cities and human settlements inclusive, safe, resilient and sustainable
Goal 3	Ensure healthy lives and promote well-being for all at all ages	Goal 12	Ensure sustainable consumption and production patterns
Goal 4	Ensure inclusive and equitable quality education and promote lifelong learning opportunities for all	Goal 13	Take urgent action to combat climate change and its effects
Goal 5	Achieve gender equality and empower all women and girls	Goal 14	Conserve and sustainably use the oceans, seas and marine resources for sustainable development
Goal 6	Ensure availability and sustainable management of water and sanitation for all	Goal 15	Protect, restore and promote sustainable use of terrestrial ecosystems, sustainably manage forests, combat desertification, and halt and reverse land degradation and biodiversity loss
Goal 7	Ensure access to affordable, reliable, sustainable and modern energy for all	Goal 16	Promote peaceful and inclusive societies for sustainable development, provide access to justice for all and build effective, accountable and inclusive institutions at all levels
Goal 8	Promote sustained, inclusive and sustainable economic growth, full and productive employment and decent work for all	Goal 17	Strengthen the means of implementation and revitalise the Global Partnership for Sustainable Development
Goal 9	Build resilient infrastructure, promote inclusive and sustainable industrialisation and foster innovation		

> **Knowledge check 27**
>
> What are the main differences between the MDGs and the SDGs?

The role of globalisation in influencing development

Globalisation refers to the world becoming more interconnected and interdependent. People around the world have become a lot more connected through developments in technology, communications and the internet. Globalisation is also evident in trade arrangements as goods and services can now be easily and quickly moved from one part of the world to another.

Many transnational companies have factories and offices in countries all over the world — for example, Ford Motors and Cadbury.

Some of the key features of globalisation are as follows.

- Globalisation has brought the world's economies closer together (especially in relation to trade and investment).
- Trade across the world has grown quickly, with many LEDCs increasing their industrial output.
- Global communications have been a key factor in the continued growth of links across the world.
- Profits made by multinational companies (MNCs) often flow back to the head offices in MEDCs and funds rarely remain in the LEDCs. However, this is changing as new financial and tax rules are introduced.
- Individual countries are less independent than they used to be. Global firms like Exxon, Shell and Walmart can have more money and more control over decisions than some governments.

Advantages of globalisation

- Globalisation is responsible for many formal employment opportunities in LEDCs.
- Often MNCs will spend a lot of money helping to improve the infrastructure and the social conditions of an area. LEDC governments will go out of their way to attract investment from big firms.
- Additional factories mean that more money is flowing into and through the country.
- Workers can receive a better education and are able to improve their skills, which makes them more employable and able to earn more money in the future.
- New skills, techniques and technologies are brought into poor countries, allowing them to develop.

Disadvantages of globalisation

- Although many LEDCs embrace any jobs that are brought into their country, these are often less well paid than similar jobs in MEDCs. This pay disparity makes LEDC workers feel less valued than those in MEDCs.
- Working conditions are not always as good as they are in MEDCs. Many factories are not owned by the MNC but are contracted out so that usual working standards do not have to be implemented and corners can be cut on hours and conditions.
- As most of the profits for anything manufactured and sold flow straight back to the headquarters of the company, little of the money stays or is invested in the LEDC partner.
- There is little job security. MNCs can be ruthless operators and will only work in partnership with a country as long as it is economically viable and safe to do so. At the first sign of difficulty a company might pull out and leave workers without a job or income.

The role of aid in influencing development

Aid involves one country or organisation giving resources to another country. This might be in the form of money, expertise (such as aid workers, healthcare professionals and/or rescue workers) or goods (water, food, blankets, tents, shelters, tools, rescue equipment and/or transport facilities).

Exam tip

This is often a difficult concept to explain in examination answers so make sure that you are well prepared to discuss the recent links formed between countries and companies.

Knowledge check 28

What is a definition of globalisation?

Different types of aid

- Bilateral aid is when help is given from one country to another country. The aid is usually tied, so that the donating MEDC can direct the money towards particular issues and priorities.
- Multilateral aid is when aid comes from world/international organisations like the WHO and the UN. Money is paid into the organisations and then used to fund projects and rescue missions around the world.
- Voluntary aid is when charities and non-governmental organisations (NGOs) are set up to help with a particular issue or support a particular place, and the public funds these. Charities like Christian Aid, Comic Relief, Oxfam and Fields of Life often provide help and workers on short- and long-term development projects.

Knowledge check 29

What is the difference between bilateral and multilateral aid?

Advantages of aid

- Aid can make a huge difference to the lives of people in LEDCs.
- Short-term and humanitarian aid can help save lives in emergency situations.
- Charities can make a big difference by working in small communities and alongside partner organisations in LEDCs.
- Aid can help to improve the standard of living of ordinary people in LEDCs.

Disadvantages of aid

- Aid does not always make a difference to the communities at which it is aimed.
- Sometimes aid does not reach the people who need it most because of corruption and poor administration systems.
- Many LEDCs have become reliant on a regular flow of aid. They become less likely to be able to stand on their own two feet without support.
- Tied aid can put strict conditions on LEDCs, which makes the aid less desirable.
- Aid can cause problems for local producers because any food aid that comes into the area might bring the food prices down and undermine any profit that local farmers might make.

Exam tip

Make sure that you have learned the different types of aid and see if you can quote examples of where the different types of aid have been applied.

Case study

The role that globalisation and aid can have in influencing an LEDC (Ghana)

The Republic of Ghana is located on the west coast of Africa, on the Gulf of Guinea. It is an LEDC, but recently its economy has been growing and it is developing at a fast rate. Ghana is a producer of petroleum and natural gas. It is also one of the largest gold and diamond producers, as well as the world's second largest producer of cocoa.

Colonialism in Ghana

Contact was first made with Europeans in the 15th century when Portuguese traders established the Portuguese Gold Coast in West Africa. The Dutch, Danish, Germans and Swedish also set up gold trading ventures along the coastline, building forts and castles. In 1874, the British established control over the area and it was called the British Gold Coast. Accra was established on the coast as the capital city. The British invested money in schools, the road and rail network, and communications.

Probably the biggest impact was the introduction of the cacao tree in 1878 as a cash crop. Within 50 years Ghana was one of the world's leading suppliers of cocoa. A famous Ghanaian saying was 'Cocoa is Ghana, Ghana is Cocoa' and this reflected the importance of cocoa to the early industrial progress of the country. In the 1920s Ghana exported some 200,000 tons of cocoa.

Ghana fought long for self-governance and became the first sub-Saharan country to gain independence from the British in 1957. However, the political journey since then has not always been easy.

Globalisation

Ghana's history and development is intertwined with its wealth of natural resources — in particular in the discovery of the country's minerals and oil. Economic reforms started in 1983 and very quickly globalisation began to have a major effect on the economy and society. In recent years the country has developed links with MNCs and with countries like China in order to continue developing trade and exports. New digital-based manufacturing plants have been built to produce tablet computers and smartphones.

MNCs have been involved in stimulating Ghana's economy and creating new jobs. Industries include mining, petroleum, telecommunications and banking with companies such as MTN, Vodafone, Guinness, Unilever and Zain. Ghana has even set up its own telecommunications MNC called Rig Communications.

One major industry that has links to many different MNCs is cocoa farming. Ghana has around 720,000 cocoa farmers and the production of cocoa is a major source of revenue. In 2008, Cadbury set up the Cadbury Cocoa partnership, aiming to invest £45 million over 10 years to help Ghana's cocoa farmers.

Ghana's Takoradi Harbour was built in 1928 and encouraged the country to develop one of the world's fastest-growing shipping industries.

In recent years the GDP in Ghana has increased dramatically — by 14% in 2011 and 8.7% in 2012. Much of this is down to the global trading and business links that Ghana has established with other countries around the world.

Ghana recently adopted an economic plan called 'Ghana Vision 2020', which aims to eventually get the country accepted as an MEDC. Before this

can happen it needs to be accepted as a newly industrialising country (NIC).

However, one major source of income in Ghana comes in the form of remittances. The UNDP estimated that there are 1 million Ghanaian migrants overseas providing Ghana with $4 billion in remittances each year.

Aid

Even though the economy in Ghana is growing faster than in all of western Europe, it still relies heavily on aid. The country receives $1 billion a year in aid, which makes up around 10% of the country's GDP.

One controversial aid project was the construction of the Akosombo Dam on the River Volta between 1961 and 1965 in order to produce cheap electricity, which resulted in the relocation of 80,000 people from 700 villages. The dam was built using bilateral aid from the UK and USA and multilateral aid from the World Bank. American firms were brought in to complete the work.

In 2010, the Ghanaian government introduced a new 'Ghana Aid Policy & Strategy', which set out some of the wider targets to help Ghana gain middle-income status. In the rationale for the policy, the government note that:

> '...since the early years of independence, Ghana has been a beneficiary of external assistance. Though external aid plays a significant role in the Ghanaian economy, it has been fairly unpredictable and/or not delivered effectively and efficiently. External aid has also not been properly coordinated and managed, and has been in most cases supply-driven...'

The specific objectives of the policy therefore were for Ghana to:

- improve country ownership and leadership of aid management processes;
- ensure effective aid coordination and management;
- manage for development results;
- strengthen mutual accountability; and
- move beyond aid dependence.

Some programmes within Ghana are trying to improve the lives of the local people. Former UN Secretary General Kofi Annan, a Ghanaian, set up the Global Fund. In August 2015, the President of Ghana signed new grants for $248 million in order to increase the number of people receiving prevention treatment for HIV/AIDS, tuberculosis and malaria.

The UK Department for International Development is involved in working in Ghana to encourage wealth creation, ensure that all children have a good education, and reduce maternal mortality rates and inequalities. Its aim is to help Ghana make improvements to how it is working towards some of the MDGs. By 2015, UK Aid (the Department's rebranded name for all its aid programmes) was able to note the following achievements:

- Over 525,000 more Ghanaians had access to family planning methods.
- 30,000 producers were able to access business services.
- 4.75 million mosquito nets were distributed in order to help prevent malaria.
- 70,000 girls were able to stay in high school through the provision of educational incentives.
- Over 9 million Ghanaians were able to vote in the 2012 national elections due to governmental support.

The Millennium Villages project was set up to demonstrate how the MDGs could be met in rural Africa through the implementation of a 10-year integrated and community-led plan. Two Millennium Village communities have been set up in Ghana. One of these, Bonsaaso, is located in the Ashanti region of Ghana, where the climate is hot and humid and surrounded by tropical rainforest. Over 80% of the land is used for the cultivation of cocoa and 70% of the 35,000 population have to survive on less than $1 a day.

The Millennium Village project so far has been able to achieve the following:
- Average crop yields have doubled (from 2.2 to 4.5 tons per hectare) due to improvements in farming methods.
- Schools that have electricity and separate toilets have been built for boys and girls.
- Malnutrition has decreased by 33% among children under the age of two.
- The majority of women now give birth in a hospital and 86% of women have at least four ante-natal care visits.
- Village members now have access to a full health facility.
- Three times as many people have access to HIV tests.
- Over 200 km of roads were repaired to improve transport links and allow farmers to sell their produce further afield.

Exam tip

This case study is a starting point for looking at how development issues might affect one country. In an exam, you need to build on this foundation with more facts and figures to support your answer, and show a thorough understanding of how each concept works.

Knowledge check 30

Describe the importance of Ghana's location in its development.

Summary

- Development is affected by a number of issues that can slow down or speed up the process.
- The UN introduced the Millennium Development Goals (MDGs) in 2000 as a means of improving global development. The eight MDGs were aimed at improving the lives of more than 1 billion people by 2015.
- The MDGs were seen as having a very wide impact on people in LEDCs.
- A new set of goals replaced the MDGs in 2015 as part of the 2030 Agenda for Sustainable Development. These were called the 17 Sustainable Development goals (SDGs or 'Global Goals').
- The SDGs are aimed at a much wider audience and are designed to help improve both LEDCs and MEDCs worldwide, to complete what the MDGs did not achieve and end all forms of poverty.
- Globalisation is a major influence on the global economy, bringing both benefits and challenges. Through globalisation, countries are becoming more connected and more interdependent.
- There are many different types of aid, including bilateral, multilateral, voluntary and tied. Aid is not always a good thing — countries can become too reliant on it.
- The case study of Ghana illustrates the main issues that affect development, especially in relation to globalisation and aid.

Emerging markets

What is an emerging market?

Trade is the flow of goods and services between producers and consumers around the world. Good trade links are necessary for any country to be able to make money. Each country achieves a balance of trade (the difference between the total cost of all imports and the value of its exports). As a result, some countries have a trade surplus — they export more than they import — and others will have a trade deficit.

LEDCs make much of their money through their exportation of primary products, which they export in a raw state to the MEDCs for processing. For this reason they are affected by price fluctuations on the raw materials. Many LEDCs do not have the technology to process the raw materials. In some cases this is because MEDCs put tariffs and charges on processed goods, thus discouraging LEDCs from carrying it out.

However, in recent years globalisation has led to LEDCs taking a bigger share of the global manufacturing industry. Electronics and textiles industries have been growing rapidly within LEDCs. A group of **newly industrialising countries** (NICs) have expanded their manufacturing. China, for example, has expanded its influence in LEDCs and now controls around 17% of all African exports and Brazil and India control an additional 9%.

An **emerging market** (or economy) is described as being a country that has some of the characteristics of an MEDC but does not meet all of the criteria. World Bank economist Antoine Van Agtmael first used the term in the 1980s. The criteria for an emerging market are as follows.

- It is a country with an economy that is progressing towards MEDC status.
- The economy will have a high annual growth rate (growth in GDP) with income between 10% and 75% of the average EU per capita income.

Newly industrialising countries Countries which are developing and are experiencing rapid industrial and economic growth.

■ This growth will be evident across a variety of economic sectors including services and manufacturing.

The BRICS and MINT nations (see below) are the most commonly noted and important emerging markets, though some economists have identified 50 countries that fall into the criteria above. India and China are considered the two largest emerging markets.

Knowledge check 31

Define the term 'emerging market'.

The growth of emerging markets

BRICS countries

The **BRICS** countries are Brazil, Russia, India, China and South Africa. These are the five main emerging markets or economies. The group was originally known as **BRIC** before South Africa joined in 2010. Each of the countries is a NIC, but they stand out among other nations as they have large, fast-growing economies and have a growing influence on global politics.

The chairman of Goldman Sachs, Jim O'Neill, first used the term 'BRIC' in 2001. The first formal meeting of the BRICS group took place in June 2009. The leadership of each of the countries meet annually at summits, the most recent having taken place in July 2015. Each of the five countries is also a G-20 member.

The BRICS countries include some of the biggest significant emerging economies around the world. In the 1990s, all were showing a GDP growth rate of around 10% or more. Each of the countries also has a very large population (India and China each have more than 1 billion people), although South Africa is significantly smaller than the others. This means that these countries have a **demographic dividend** (a stage of development when the fertility level in a country is starting to fall but there is still a young working age population that will continue to boost the economy).

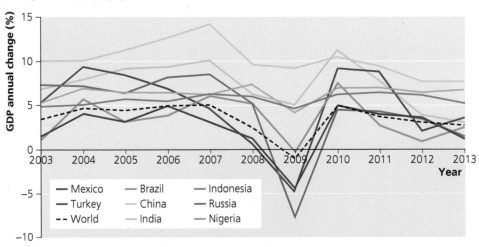

Figure 20 GDP growth in BRIC and MINT countries (% annual change 2003–2013)

Table 11 BRICS countries: key facts

	Brazil	Russia	India	China	South Africa
GDP growth rate (2015)	-3.5%	-2.7%	7.5%	6.8%	1%
GDP per capita (US$) (2015)	8,802	8,447	1,688	8,280	5,784
Population (millions)	204	144	1,276	1,376	55
Exports ($ billion)	396	542	462	2,021	101
Imports ($ billion)	279	358	500	1,780	107
Literacy rate	95.5%	99.9%	74.0%	95.1%	93%
Life expectancy (years)	74.5	70.1	68	75.8	57.4

In 2013, the BRICS countries started to plan for the development of the New Development Bank (formerly the BRICS Development Bank). This was a new bank with a currency reserve of $100 billion. The aim of the bank was to provide funding for infrastructure and sustainable development projects in BRICS and other developing countries. Each country has an equal share of ownership in the bank.

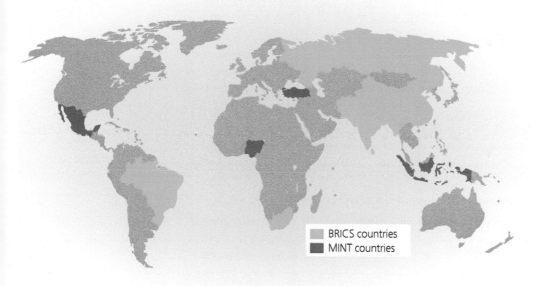

BRICS countries
MINT countries

Figure 21 The location of the BRICS and MINT countries

MINT countries

MINT countries are identified as **M**exico, **I**ndonesia, **N**igeria and **T**urkey. The financial corporation Fidelity Investments coined the term, but Jim O'Neill made it a more common term. The term was first used to denote economic and financial growth. Initially the group was to include South Korea instead of Nigeria (MIST) countries. However, O'Neill noted that 'Mexico, Indonesia, Nigeria and Turkey all have very favourable demographics for at least the next 20 years, and their economic prospects are interesting'.

Each of the countries was identified as having particular economic strengths that showed it to be resilient within the global markets (see Figure 21).

Table 12 MINT countries: key facts

	Mexico	Indonesia	Nigeria	Turkey
GDP growth rate (2010–13)	3.4%	6.0%	6.8%	5.9%
GDP per capita (US$) (2015)	10,230	3,491	3,203	10,529
Population (millions)	125	254	177	76
Exports ($ billion)	427	201	91	223
Imports ($ billion)	477	231	80	271
Literacy rate	93.9%	93.9%	61.3%	98.2%
Life expectancy (years)	77.14	70.61	52.1	75.2

Mexico has the 15th largest global economy. It is a member of the North American Free Trade Association (NAFTA) and the Organisation for Economic Co-operation and Development (OECD). These are both very important organisations and have aided a recent rise in GDP. A large local wealth has been created, which has developed a large domestic consumer market that has helped to create further jobs.

Indonesia has benefited greatly from the global shift of manufacturing industries from MEDCs to LEDCs. It is ranked seventh in terms of global GDP and has a fast-growing economy. It is the largest economy in Southeast Asia and is also a member of the G-20 major economies. It is the fourth most populated country on the planet.

Nigeria has rapidly become important due to its role as a major oil exporter. It is ranked 20th in the world in relation to GDP and is the largest economy in Africa. Many economists note that the manufacturing sector is very big but is still underperforming. Internal ethnic/political/religious issues are seen as holding back further economic growth.

Turkey has a strategically important position in the world and is often seen as 'a bridge nation between the West and the Islamic world'. In 2010 its economy grew at a rate of 9.2%, making it the fastest-growing economy in the world. It has the 18th largest economy in terms of GDP but is expected to continue to grow at a fast rate over the next 20 years.

Controversies within the emerging markets

- The BRICS countries are a self-contained group but the MINT countries are just a loose group of like-minded countries without a central organisation to combine them.
- The BRICS countries seem to have an economic and political agenda to challenge the G-20 and G7 countries and their economic dominance over the rest of the world.
- Some of the countries (notably Nigeria and Indonesia) have struggled with corruption issues in the past.
- Turkey, China and Russia are not known for their good record in human rights and this is sometimes seen as being a major roadblock to further development.
- Argentina, Indonesia, Egypt, Nigeria, Sudan, Syria, Bangladesh and Greece have all expressed an interest in joining the BRICS group of countries. It will be interesting to see how new groups of countries can continue to work together in the years to come.

Knowledge check 32

What are the main differences between BRICS and MINT countries?

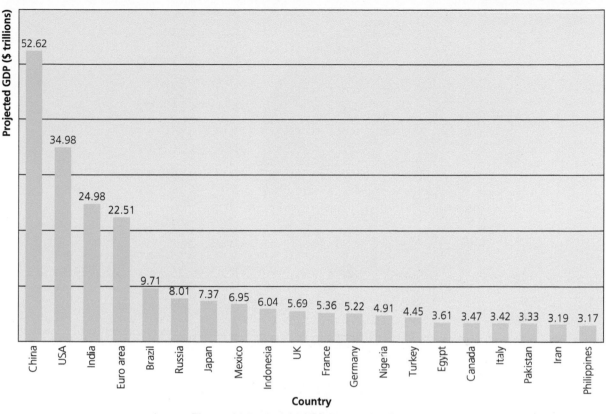

Figure 22 Projected GDP figures for 2050

Case study

Emerging markets in a BRICS country

Some recent articles on global economics have noted that many of the BRICS economies seem to have stalled or even gone into a decline over the past few years. Figure 22 shows how China's dominance is expected to grow over the rest of the world. However, India and Brazil are also expected to grow quickly in the long term, overtaking many of the older, more established MEDC countries.

Emerging markets in India

Economists note that out of the BRICS countries, India has one of the fastest-growing economies. In 2015, a *Fortune* magazine article noted that

India was 'the lone BRICS country that's worth holding on to'.

India is often described as being the world's largest democracy and is home to over 1 billion people. The GDP from 2000 to 2011 averaged 7.5%, hitting a high of 11.8% in 2003 and a low of 1.6% in 2002. Growth in India has remained strong in recent years with estimates of a 7.8% growth rate in 2016.

Some of this high performance has come about as a result of recent falls in the price of crude oil and natural gas processing, which reduces India's massive energy bill.

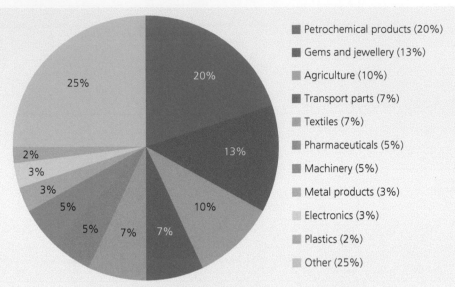

Petrochemical products (20%)

Gems and jewellery (13%)

Agriculture (10%)

Transport parts (7%)

Textiles (7%)

Pharmaceuticals (5%)

Machinery (5%)

Metal products (3%)

Electronics (3%)

Plastics (2%)

Other (25%)

Figure 23 The value of exports from India (2014)

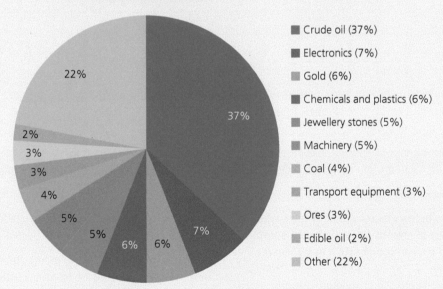

Crude oil (37%)

Electronics (7%)

Gold (6%)

Chemicals and plastics (6%)

Jewellery stones (5%)

Machinery (5%)

Coal (4%)

Transport equipment (3%)

Ores (3%)

Edible oil (2%)

Other (22%)

Figure 24 The value of imports into India (2014)

Global investment firms make much of the political stability in India, which allows any potential multinational company to explore ever-expanding investment opportunities. Within India there have been big changes to the jobs that people are doing. There is a rise in the professional and service sector and a decline in manufacturing (manufacturing sector activity fell to its lowest level in 2 years in 2014). The service sector contributed 57% of its GDP in 2013. IT services have grown particularly quickly — the large number of English-speaking workers can be used in IT outsourcing (e.g. call centres) and software services. India's agricultural economy is becoming increasingly diverse and ranges from traditional village farming to modern agriculture, making up 17% of the Indian economy. Industry makes up 26% of the GDP — growth areas such as automobile manufacturing have made India's auto-industry one of the largest in the world, with an annual production of 21.48 million vehicles. The chemical industry is also another important component (making up 5% of the GDP). India supports many oil refineries and petrochemical operations. Engineering and pharmaceuticals are currently big growth areas within the Indian economy and are set to continue to expand over the next 10 years.

One success story that highlights the emerging growth in India can be seen in the recent launch of the Freedom 251 3G smartphone. The smartphone only costs 251 rupees, or £2.60. Indian company Ringing Bells manufactures the phone on the Google Android operating system. Within two days of release, over 30,000 units were sold and the official website through which the phone could be purchased collapsed for 24 hours.

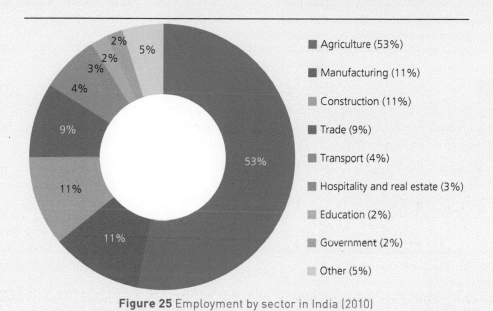

- Agriculture (53%)
- Manufacturing (11%)
- Construction (11%)
- Trade (9%)
- Transport (4%)
- Hospitality and real estate (3%)
- Education (2%)
- Government (2%)
- Other (5%)

Figure 25 Employment by sector in India (2010)

The massive growth of the Indian economy in recent years is down to a number of factors.

- **Economic liberalisation:** until 1990, the Soviet Union was India's main trading partner. The fall of Communism across Europe caused an economic crisis in India, which led to an economic liberalisation in 1991.
- **High standard of living per capita income:** the number of 'middle class' people in India has been increasing quickly (over 267 million people by 2016).
- Development of medical facilities and infrastructure.
- Increase in foreign investment.

Some of the challenges facing the Indian economy over the next few years, which will need to be overcome in order to continue sustained growth, include the following.

- **Inflation:** as wages, property prices and food prices all start to rise, inflation is going to be an ever-increasing issue in India. Inflation hit 9.6% in June 2011 but has since reduced to 3.78% (2014).
- **Low educational standards:** improvements in education are often seen as one of the biggest factors in fuelling the continued economic development of the country. Adult literacy rates in India stand at 74% in 2011. India has made excellent progress in increasing primary education attendance (up to 93% in 2011). A total of 69% of children will attend secondary school and 25% will stay on to university and tertiary education. There remains a significant contrast between attainment in urban and rural areas. Some sources note that over 50% of women in India are still illiterate.
- **Lack of infrastructure:** many rural poor people still lack the most basic amenities, such as running water, and access to electricity and sewage services. In 2012 India was ranked 89th out of 142 countries for its infrastructure. The *World Economic Forum Global Competiveness*

Report noted that: 'the Indian business community continues to cite infrastructure as the single biggest hindrance to doing business in the country.' Investment is encouraged in the rail and road networks in particular.

■ **High levels of debt:** over the last 10 years a property boom in India has increased the amount of lending in the country (a growth of around 30%). Much of this lending is based on assumptions that the economy will continue to grow at a steady rate — any decrease or pause in the growth rate or an increase in the interest rates could bring debt problems.

■ **Inequality on the increase:** it was hoped that the economic growth of the last 10 years would help the most vulnerable in Indian society. This has not been the case, and in many ways the gap between the richest and the poorest in Indian society has increased, not decreased.

Exam tip

India is only one example of a country that has been emerging in recent years. You might want to do some research into a second country — but do make sure that it is taken from the list of BRICS and MINT countries.

Summary

■ Trade is the flow of goods and services between producers and consumers around the world.

■ Many LEDCs make their money through the exportation of primary products, which they export in a raw state to the MEDCs for processing.

■ An emerging market or economy is a country that has some of the characteristics of an MEDC but does not meet all of the criteria.

■ The BRICS countries are Brazil, Russia, India, China and South Africa. They are seen as standing out among other nations as they have large, fast-growing economies and have a growing influence on global politics.

■ The MINT countries are Mexico, Indonesia, Nigeria and Turkey. Jim O'Neill identified each of these countries as having particular economic strengths that showed them to be resilient within the global markets.

■ One case study of a BRICS country that has been doing well in recent years is India. The economy of India is continuing to grow at over 7% GDP per year, which makes the Indian economy one of the fastest growing on Earth.

Questions & Answers

The AS Unit 2 Geography paper includes six questions:

	Compulsory?	Marks (out of 75)	Exam timing (out of 75 minutes)
Section A			
Q1 Population: short structured questions	Yes	15	15
Q2 Settlement: short structured questions	Yes	15	15
Q3 Development: short structured questions	Yes	15	15
Section B			
Q4 Population: extended question	Answer **two** from questions 4, 5 or 6	30 marks = 15 marks for each question	30 minutes = 15 minutes for each question
Q5 Settlement: extended question			
Q6 Development: extended question			

Examination skills

As with all AS exams there is little room for error if you want to get the best grade. Gaining a grade A is not easy in AS geography so you need to ensure that every mark counts.

The examination papers for AS Unit 1 and Unit 2 are both 1 hour and 15 minutes. There are 75 marks available on each, which means that you get 1 mark per minute to work your way through the paper. The main reason why so many students struggle with this paper is that they fail to manage their time appropriately and as a consequence they do not have enough time left to answer the essays at the end in sufficient detail. If you find that you have time left over in this exam, the chances are that you have done something wrong.

Exam technique

Students often find it difficult to break an exam question down into its component parts. On CCEA exam papers, the questions are often long and difficult to understand, so you need to work out what the question is asking before you move forward.

Command words

To break down the question properly, get into the habit of reading the question at least *three* times. When you do this it is sometimes a good idea to put a circle round any command or key words that are being used in the question.

A common mistake is failing to understand the task being set by the question. There is a huge difference between an answer asking for a discussion and one asking for an evaluation. The main command words used in the exam are as follows.

- **Compare:** what are the main differences and similarities?
- **Contrast:** what are the main differences?
- **Define:** state the meaning (definition) of the term.
- **Describe:** use details to show the shape/pattern of a resource. What does it look like? What are the highs, lows and averages?
- **Discuss:** describe and explain. Argue a particular point and perhaps put both sides of this argument (agree and disagree).
- **Explain:** give reasons why a pattern/feature exists, using geographical knowledge.
- **Evaluate:** look at the positive and negative points of a particular strategy or theory.
- **Identify:** choose or select.

Structure your answer carefully

Sometimes the longer questions on exam papers — for example, short structured questions worth 6 marks or extended writing questions worth 15 marks — can be an obstacle for students. Later in this section we will look at some questions and give more guidance about how you should structure your answers.

One simple approach to consider is drawing up a brief plan for your answer so that you know where it is going and how you will cover all of the main aspects of the question. For example, you could draw a box to illustrate each element needed within an answer and fill each one with facts and figures to support the answer, using the marking guidance to help you work out how much time to spend on each section (examples are shown for the essay questions below).

Show your depth of knowledge of a particular place/case study

The extended writing essay questions on the exam paper are usually focused on giving the student the opportunity to apply knowledge and understanding of case study material to a particular question. It is really important to show what you know here.

Examiners are looking for specific and appropriate details, facts and figures to support your case. The better you know and understand your case studies, the higher the marks you can potentially achieve.

About this section

A practice test paper with exemplar answers is provided. This will help you to understand how to construct your answers in order to achieve the highest possible marks.

Examiner comments

Some questions are followed by brief guidance on how to approach the question (shown by the icon ⓔ). Student responses are followed by examiner's comments. These are preceded by the icon ⓔ and indicate where credit is due. In the weaker answers, they also point out areas for improvement, specific problems, and common errors such as lack of clarity, weak or non-existent development, irrelevance, misinterpretation of the question and mistaken meanings of terms.

Section A

Question 1 Population: short structured questions

(a) Describe **one** problem with population data collection in an LEDC and **one** problem associated with population data collection in an MEDC. (4 marks)

e 1 mark is awarded for the identification of a problem with data collection in an LEDC and 1 mark for more a more detailed description that includes valid facts and figures to support the answer. Max 2 marks for LEDC.

1 mark is awarded for the identification of a problem with data collection in an MEDC and 1 mark for a more detailed description that includes valid facts and figures to support the answer. Max 2 marks for MEDC.

Student answer

(a) In LEDCs a lot more people in the country cannot read and write and this makes it difficult to collect accurate information. They cannot read the questions themselves and need more help, and they might not even be aware of when they were born.

In MEDCs the biggest problem is making sure that information is filled in accurately. For example, some students might live at university but they might be counted in a house in the city or at their parents' home. Some people will be counted twice while others will deliberately not put their name on the form in case they have to pay more tax.

e **4/4 marks awarded.** Not all the information in the answer is totally relevant but there is some good depth on the problems in LEDCs, and some balance in looking at issues of accuracy in MEDCs.

A range of problems could be discussed here. The answer needs to have balance, describing one problem in LEDCs and one in MEDCs. There is no need for comparison between the two problems. Answers might include the size of countries, language and access problems, gender issues, funding and training problems and identity fraud.

(b) Define the 'epidemiological transition' and describe how it varies globally. (6 marks)

e An answer here should seek understanding for the definition (3 marks). The spatial element of the question requires candidates to look at the contrast between MEDCs and LEDCs. Any answer should focus on the ways that MEDCs have made their way through the process and this should be compared with the slower progress of the LEDCs. There is no real requirement for examples or case studies. If there is no spatial variation, maximum 1 mark out of 3. (2 parts x 3 marks each)

> **(b)** The epidemiological transition is a theory linked to the Demographic Transition Model, which looked at how mortality rates in different countries were changing and that the main causes of death would go from being infectious diseases like small pox to more chronic diseases and diseases associated with longer lives like cancer, stroke and heart disease.
>
> Rich countries like the UK and France find that most of the people will live long lives and die over the age of 65 from things like cancer, stroke and heart disease.

🅔 **4/6 marks awarded.** The first part of the answer that deals with the definition of the epidemiological transition is well answered with appropriate detail, and gains the candidate the full 3 marks available for this section. However, there is little development of the second part of the question and it only mentions MEDCs with some repetition of information, so 1 mark out of 3.

The answer should go further to explain the different stages: Stage 1 the age of pestilence and famine that is still dominant in a number of LEDCs, the age of receding pandemics and then the age of degenerative and man-made diseases that is usually found in MEDCs.

(c) Describe how Ester Boserup's thoughts on population sustainability differ from those of Thomas Malthus.

(5 marks)

🅔 **Level 3 (5 marks):** a very good and effective answer that is packed with detail and shows a good understanding of the differences between the two theories.

Level 2 (3–4 marks): a more rounded answer that will have detail on both Boserup and Malthus but might lack the depth required for Level 3.

Level 1 (1–2 marks): a basic answer, which lacks explicit reference to detail and does not give specific examples or reference to either theory.

> **(c)** Ester Boserup was a Danish agricultural economist who wrote a response to some of the gloomy ideas on population sustainability put forward by Thomas Malthus. She believed that when the population growth hit a crisis point, that the people would be able to innovate to make sure that the balance between resources and population was maintained.
>
> Thomas Malthus is the father of demography and he was a pessimist who thought that when the population hit the crisis point that this would cause a lot of problems and people would die from famine and starvation. He told people not to have children. He had some positive and negative checks on the population.

🅔 **3/5 marks awarded.** This is a Level 2 answer and while there is some good information and depth in relation to Boserup, the information and depth on Malthus is not as good and there is some irrelevant information for the argument.

The task in this answer is to show understanding of the two different theories, so there should be good depth shown for both Boserup and Malthus. This question does not ask for an evaluation though this is a possible question that could come up in relation to this material.

Question 2 Settlement: short structured questions

(a) Study the diagram below, which shows the percentage of the world's population living in urban settlements in 2010. Describe and explain the distribution of areas around the world that have high percentages of urban and rural populations.

(6 marks)

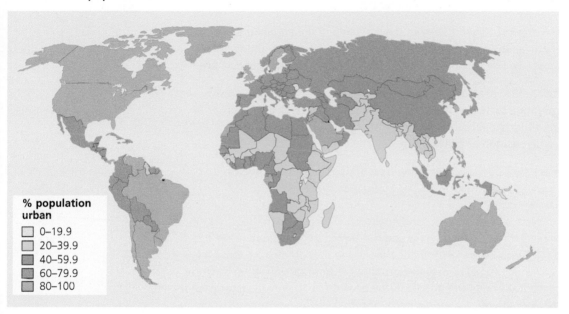

% population urban
- ☐ 0–19.9
- ☐ 20–39.9
- ☐ 40–59.9
- ☐ 60–79.9
- ☐ 80–100

Percentage of the world's population living in urban settlements, 2010

Student answer

(a) From the map I can see that in places like USA, Canada, Brazil and Saudi Arabia there is a high percentage of people (80–100%) who live in cities. There are also quite a lot of people in the western European countries (60–79.9%). In contrast to this, many of the countries across Africa and Asia have a low percentage of people living in urban areas. Many of the East African countries and India and Pakistan have only 20–39.9% of people living in cities. The reason for this is that the countries with many people living in cities seem to be the richer/MEDC countries. People want to live in a city as this is where all the jobs are, and people know that living in a city will give them better lives.

ⓔ 5/6 marks awarded. This begins with a good description of the distribution of urban and rural areas, so 3 marks are awarded. However, the explanation is a little weaker and needs more depth for full marks.

A description of the pattern of the distribution is required before attempting to explain this pattern. 3 marks are awarded for description. The places that have a high percentage of urban population (80–100%) are the USA, Canada, Argentina, Brazil and Australia. Credit can be given for locations using continental titles. The places with a low percentage of urban population are generally found in the 20–39.9% category and are in Africa and Asia. This is where people still largely live in rural areas. The other 3 marks are given for an explanation of how most of the urban areas are in rich/MEDC countries and the rural areas are found in many of the poorer/LEDC countries.

(b) With reference to your case study, discuss the social and economic deprivation in an inner city area of an MEDC. (6 marks)

ⓔ **Level 3 (5–6 marks):** A good all-round discussion of the range of social and economic indicators, and figures used to develop the discussion of deprivation in a particular case study. Specific place names are mentioned in the case study.

Level 2 (3–4 marks): Some good content but in less depth. Figures linked to indicators might be missing. There might be inaccuracies or omission of discussion of either social or economic factors. There might not be a comprehensive discussion of specific places.

Level 1 (1–2 marks): A poor answer showing a limited understanding of the question.

(b) In Belfast there is a lot of social and economic deprivation in the inner city. Parts of east Belfast, in Sydenham and the Newtownards Road, have been suffering for years because of the decline of the manufacturing industry in this area. This has led to massive amounts of unemployment in the area — currently 14% of the people in the area do not have jobs (compared with the NI average of 6%). This shows that this area is hit more during the recession. Many workers do not have a good education, which makes it difficult for them to get a better job.

In terms of social deprivation, this area of Belfast has one of the highest percentages of free school meals (21%), which shows that the government needs to support the families and make sure that they get good food. Few students leave school with more than five GCSEs (about 24%).

ⓔ **5/6 marks awarded.** This is a good discussion of the key ideas about social and economic deprivation in the city. The discussion about economic deprivation is probably better developed than that about social deprivation.

This question is looking for discussion of the social and economic deprivation in your MEDC case study city. Any answer that leaves out either social or economic deprivation will be limited to 3 marks. There should be a discussion of the different levels of deprivation, with some mention of both areas and specific locations in the city, and reference to a range of indicators.

(c) Briefly describe the difference between AONBs and ASSIs. (3 marks)

e This question is looking for a good understanding of the two measures — the differences can be very subtle between the two.

1 mark is awarded for a basic answer that might only relate to one of the measures.

2 marks are awarded for a more detailed answer that shows some understanding of both measures.

3 marks are awarded for an answer that has a clear understanding of the subtle differences between the two measures.

(c) AONB stands for Area of Outstanding Natural Beauty. This is an area of the countryside that is considered to have landscape of good value. The idea is that the area will be protected and conserved and that planning laws would stop the development of things that might damage the environment. ASSI is an Area of Special Scientific Interest. This is used to protect rocks or biological wonders and some of the most scientifically unique places across the UK.

e **3/3 marks awarded.** This is a good discussion of the key attributes of the two measures used to manage the countryside.

Question 3 Development: short structured questions

(a) Distinguish between globalisation and trade. (3 marks)

Student answer

(a) Globalisation is the process whereby countries have been able to develop links so that they are effectively closer to each other. The transport links and communications between countries have brought them closer together, so that some companies like Nike and other NMCs have offices all over the world. Trade is the movement of goods from one place to another as imports and exports.

e **3/3 marks awarded.** This is a good explanation of globalisation, which goes on to give a definition of trade, showing the differences between the two concepts.

Globalisation is the process whereby the world is becoming more interconnected and interdependent. It is caused by the movement of money and capital from one place to another. Trade is the flow of goods and services between producers and consumers around the world. There is overlap between the two concepts but discussion here needs to highlight the difference. If one aspect only is discussed, no marks can be awarded.

(b) Identify and describe **one** advantage and **one** disadvantage of aid. (6 marks)

> **(b)** Aid can sometimes be a bad thing for a country as it encourages corruption. When money is given to a country for a project, sometimes the funds can be redirected into the bank accounts of government officials. The money does not reach the people who really need it.
>
> On the other hand, aid is also a very useful support for people as it can bring food in a crisis.

ⓔ 4/6 marks awarded. This identifies one valid issue and goes on to describe this in some depth, noting how corruption means that people might not get to see any of the aid given — the disadvantage is well covered but the second part of the answer on the advantage of aid is not well answered and needs a lot more detail.

Aid is the process whereby one country or organisation gives resources to another country. Usually aid is a positive, but the answer needs to focus on a brief discussion of one disadvantage. 1 mark is awarded for identifying a disadvantage and 2 further marks are for describing this in depth. So 3 marks are awarded for detail on the disadvantages of aid and 3 marks for detail on the advantages of aid.

(c) Describe and evaluate the impact of **one** of the Millennium Development Goals (MDGs). (6 marks)

ⓔ Level 3 (5–6 marks): Description and evaluation of one MDG clearly made with supporting details of the specific targets and a balanced treatment of the positive and negative impacts of the goal. A final concluding statement will be made.

Level 2 (3–4 marks): A good answer that is accurate but there is less factual detail or no use of figures. Evaluation might be weak or non-existent.

Level 1 (1–2 marks): Answer lacks depth or detail.

> **(c)** Goal 1 of the Millennium Development Goals is concerned with eradicating extreme poverty and hunger. Some of the main targets involved in this goal were to try and halve the number of people who earned less than $1 a day. Also, there was an aim to have full employment for all by 2015 and to reduce by half the proportion of people who suffer from hunger. This goal has had some success since 1990. The number of people who live in extreme poverty has gone down from 1.9 billion in 1990 to 836 million in 2015. The number of people who are malnourished has gone down from 23% to 13%. However, it is not all good news — the recent global economic crisis has caused job losses, with young people in particular bearing the brunt of the crisis. In addition, even though poverty is down, over 1.2 million people are still living in extreme poverty globally. Therefore, this MDG has had some measure of success but there is still a lot of room for improvement.

ⓔ 5/6 marks awarded. The answer goes into some good detail and clearly describes some of the targets that are within this particular development goal. There is an attempt to evaluate the impact of the goal and there is some good positive and negative impact but the discussion could have gone into a little more depth to get a top Level 3 mark.

Reference needs to be made to description of the MDG, followed by a balanced evaluation and a final concluding statement (for top Level 3).

■ Section B

Question 4 Population: extended question

Describe and explain the operation of the national fertility policy you have studied and show how it responded to a population resource imbalance.

(15 marks)

ⓔ **Level 3 (11–15 marks):** Any answer here will be full and detailed, for example in relation to China, facts about the 'Granny Police' will be common. Clear information needs to be presented on how the policy was introduced as a response to population resource imbalance.

Level 2 (6–10 marks): A good answer that might only deal with one aspect of the questions or where a full answer is attempted but might require more depth and detail linked to the case study.

Level 1 (1–5 marks): A limited answer that lacks depth. There might be some basic description but the answer lacks explanation and detail. The case study might be inappropriate or incorrect.

Student answer

In 1979, China introduced a 'One-Child Policy', as part of their family planning programme, which denied freedom of choice in terms of fertility. This policy particularly effected urban populations or 35.9% of the population. Methods of upholding this policy have ranged from fines relating to the parents' income to forced abortions and sterilisations.

The aim of the policy was to reduce the number of births through the use of all means. This meant that the state kept a very careful watch on the young female population and set up measures to check that they were only having one child (including the 'Granny Police'). If a couple did have a second child, fines would be given to the family and it could result in members of the family losing their jobs or their houses. Many forced abortions were carried out and sometimes a woman was also subjected to a sterilisation process.

The reason as to why such a policy was introduced is vast, and incorporates economic and social factors. In the early years of Mao Zedong's Communist leadership of China, the government encouraged families to have many children, under Mao's belief that a larger population would lead to a more productive nation. However, infant mortality rates declined from 227 to 52 per 1,000 births, which caused the life expectancy to rise to about 62 years and people to live longer. As a result, the population increased exponentially, from 540 million to 940 million between 1949 and 1976. In order to provide population sustainability, the population must not exceed the limit where the resources can sustain it. It was suggested that China could only sustain a maximum of approximately 800 million people, which led to the fear of a Malthusian Crisis. One such example of this was during China's 'Great Leap Forward' (1958–61) where the country attempted to increase industrialisation at the expense of agriculture, decreasing food supplies. This resulted in the great Chinese famine, which led to the deaths of 13–30 million civilians, which may be identified as a 'Malthusian Check'. As a result, to slow population growth, the One-Child Policy would greatly reduce potential births and thus slow down population growth, despite the potential of it increasing to 1.5 billion in 2000. As well as reducing the demand for natural resources, with the intention of restoring balance, the One-Child Policy was a method of preventing a financial crisis by steadying the labour rate and reducing unemployment, which may otherwise increase. Not only that, but with only one child, many parents would be able to have a greater disposable income as money does not have to be divided among a large group of children, which would also greatly boost the economy. There is evidence of this being successful, as the GDP increased from $800 to $6,500 for the duration of the One-Child Policy.

ⓔ 11/15 marks awarded. This is a low Level 3 answer. There is some discussion of the different aspects of the operation of the policy but the answer could go into a lot more depth. The discussion of the reasons behind the introduction of the One-Child Policy in China in the first place as a response to the population resource imbalance is well done and has some good depth.

A good way of organising this answer might be to use the 'planning boxes' technique. Students could structure their answer/plan in the following way and allocate time accordingly:

Question 4: Describe and explain the operation of the national fertility policy you have studied and show how it responded to a population resource imbalance. (15 marks)

Describe/explain the operation of the fertility policy	Show how this policy responded to a population resource imbalance
How did the fertility policy work? ■ Mechanics of the OCP (China) (4 mins) ■ Why was the fertility policy organised like this? (4 mins)	What led to the OCP being introduced in the first place? (7 mins) ■ Great Leap Forward ■ Pro-natalist policy and famines

Question 5 Settlement: extended question

Describe and explain some of the issues of the inner city in MEDCs in one city that you have studied. (15 marks)

ⓔ Answers need to name an MEDC city (e.g. Belfast) and then the answer needs to address some of the issues facing the inner city in MEDCs: economic and social deprivation, re-urbanisation and gentrification. At least two issues should be addressed.

Level 3 (11–15 marks): A detailed and well-written answer will cover at least two of the MEDC inner city issues. Answer will show good depth and understanding from the case study, with place names and figures where appropriate.

Students should be aware that the exam board expects students who choose to discuss economic and social deprivation to go into depth in describing clearly the levels of deprivation, using a range of social and economic indicators.

Level 2 (6–10 marks): Answer is still good but the depth of knowledge in relation to the case study may be less. Answer might have addressed one issue in good depth but another issue at a sub-optimal level. Answer might lack figures and case study details.

Level 1 (1–5 marks): Answer lacks depth in relation to the case study or might show a limited knowledge of the main issues in the MEDC city. Quality of written communication is poor and/or limited.

Student answer

Belfast is a great example of a city that has gone through some urban development and gentrification programmes over the years. Belfast began as a small group of houses sited at a fording place over the River Lagan at the head of Belfast Lough. This then became an important crossing point on a route that linked counties Antrim and Down. The population continued to increase and the fastest period of expansion was in the 18th century when factories began to be built in the town, and then in 1830 the first steam powered machinery was introduced so that during the next century the town grew to become the most important manufacturing town in Ireland. People were attracted by employment opportunities in textiles, shipbuilding and engineering.

In the 1950s programmes of slum clearances and redevelopment began, although they did not really become apparent until the 1970s. The urban population continued to grow and the city was like most industrial cities as it became more and more congested, and government strategies came into place where they created 'satellite' towns such as Craigavon and Antrim, to which some of the population was moved.

This eased the pressure on the Belfast metropolitan area. The Northern Ireland Multiple Deprivation Report states that some of the areas that have the highest deprivation in NI are in inner city areas, such as the Falls or the Shankill. This report was generated using 52 separate indicators of deprivation. Some of these include income, employment, education and health and disability. The unemployment rate in Northern Ireland is 5.4% and in Belfast the rate is 7.4% and in west Belfast's Shankill it is 13.5%. In 2012/13, 40% of students attending secondary schools received free school meals whereas in grammar schools it was only 7.4%. School attendance is another thing that is recorded and absenteeism in the Falls was recorded at 8.8% whereas the Shankill is at 11.2%. A final measurement of deprivation is the proportion of working-age adults who have no qualifications; in north Belfast's Water Works 32% of the population has no qualifications. This deprivation arises because people living in these areas are in a cycle of poverty and it is difficult to break free from this cycle, as the disadvantaged children are more likely to fail at school as they have a high percentage of absenteeism, which means that they will miss out on a lot of things at school, they will leave with very few qualifications and have little motivation to find a job, therefore they are less likely to get a job, which has an effect on their community as less money is being spent in their local shops.

In the 1980s and early 1990s there was a Laganside redevelopment as the Laganside Corporation became responsible for developing 140 hectares of land alongside the River Lagan and 70 hectares in the Cathedral Quarter. By the time the project was completed it was estimated that over 14,000 permanent jobs had been created. A total investment of £939 million resulted in the creation of 213,000 m^2 of office space, 83,000 m^2 of retail and leisure space and 741 housing units. This brought a lot of new interest and new opportunities into Belfast, however many of the communities who lived beside the Laganside development felt that that they had been ignored and felt no positive impact from the regeneration.

Belfast continues to develop new infrastructure and fund new and successful projects, such as the Titanic Quarter, however there are still high percentages of unemployment and many people are on benefits, although the government is trying to address these issues and get people 'Backin' Belfast' (as the 2013 social media campaign of the same name — designed to drive footfall back into the city and restore confidence in local businesses and increase civic pride — endeavoured to do).

ⓔ **11/15 marks awarded.** This low Level 3 answer is comprehensive and shows a command of the case study as well as application of knowledge to the question. However, much of the first paragraph is not actually addressing the question and would only be wasting time in the exam. The main part of the answer to score points starts in the second paragraph. The information in relation to economic and social deprivation is very good and combines facts and figures with a knowledge of the local places.

The second issue of redevelopment is not treated in the same amount of detail and this is why the answer does not score more highly into Level 3. A lot more specific depth would be needed to take this into a high Level 3 answer. For the question:

Describe and explain some of the issues of the inner city in MEDCs in one city that you have studied. (15 marks)

You could use the 'planning box' technique in the following way.

Issue 1: Economic and social deprivation	Issue 2: Re-urbanisation
■ Describe the main measures of deprivation (economic and social data) (4 mins) ■ Explain how these create issues in the city — how does poverty and unemployment affect the people in the city? (4 mins)	■ Describe the main features of re-urbanisation (e.g. Laganside) (4 mins) ■ Explain how this re-urbanisation programme creates issues for the city — jobs/congestion/building work etc. (3 mins)

Question 6 Development: extended question

Globalisation and aid have had a big impact on the development of LEDCs. Define these issues and show how they have affected the development of LEDCs in a positive and/or negative way.

(15 marks)

e Globalisation and aid are both issues that can have a big impact on the development of countries. The answer requires a definition of each of the issues and then a deeper explanation of how each process has played a role in affecting the development of LEDCs. A maximum of Level 2 will be awarded if no definition of each process is given.

Level 3 (11–15 marks): A detailed account that contains a solid understanding of the main components of both processes. Clear understanding is shown of the positive and/or negative effects that each process brings to LEDCs or to particular named LEDCs. Good use of terms to support the answer.

Level 2 (6–10 marks): Some good depth in the answer but level of detail might be less. Students who leave out one process completely will be limited to this level.

Level 1 (1–5 marks): Answer might lack knowledge of the issues under discussion. Answers that focus only on the definition but have no elaboration of the effect on development will be limited to this level.

Student answer

Globalisation has had a big impact on the development of LEDCs. It is the process where the different countries become more connected and dependent on each other. This is mostly caused by the increase in trade and uses much of the new technology and communications to ensure that products can be moved from one part of the world to another.

Globalisation can have a big impact on countries as this often brings much needed hard cash into poor LEDCs. Many multinational companies set up factories and offices all over the world — they want to be able to make products cheaply and then transport them for sale to the MEDCs. Globalisation brings jobs to people in the LEDCs and this provides formal jobs, training and a stable source of income. This money will then filter through the whole economy. Sometimes the MNCs will also spend money helping the LEDC to improve its transport infrastructure — there will be investment in ports and motorways. Workers will be able to get new skills and become better educated.

Aid is a very complex thing as it is usually when one country is giving resources or help to another country. There are usually seen to be three different types of aid. Voluntary aid is when charities collect up money to respond to different issues within a country. For example, Sport Relief and Comic Relief collect money to help people in other countries and at home. Multilateral aid is when help is given from global organisations like the UN or the WHO. Bilateral aid is the most common and is when help is given from one country to another.

ⓔ 8/15 marks awarded. This does include definitions of the two different concepts — globalisation and aid. However, the explanation of how these two ideas have affected the development of LEDCs could have been taken further. The discussion on globalisation is good but lacks specific details about a country or countries. The discussion of aid is a lot less detailed and although it does talk through the different types of aid, there needs to be a lot more specific depth and discussion of the ways that aid has had a positive or negative impact on people.

For the question:

Globalisation and aid have had a big impact on the development of LEDCs. Define these issues and show how they have affected the development of LEDCs in a positive and/or negative way.

(15 marks)

You could use the 'planning box' technique in the following way.

Globalisation	Aid

■ Definition: what is globalisation? (3 mins)
■ How has globalisation affected development in LEDCs? (4 mins)
  – Positive impact
  – Possibly negative impact
	■ Definition: what is aid? (4 mins)
	■ How has aid affected development in LEDCs? (4 mins)
	  – Positive impact
	  – Possibly negative impact

Knowledge check answers

1 The census helps government departments to plan for the next 10 years, looking at key trends so that decisions can be made in relation to the size of population, education, health and disability, housing, employment, ethnic groups and transport.

2 Reliability issues for vital registration in an MEDC might be that people do not follow the correct procedures to get their children registered or they might decide they do not want their children to be registered. Although the system is relatively simple it can be easily avoided.

3 The main issues of census reliability in MEDCs are usually down to human error. Mistakes can be made in making sure that all areas are covered and that accurate maps/ housing areas are drawn up. Some people are wary about what the government will use the information for, and they might supply incorrect information. Diverse language needs and special needs might also mean that coverage is not universal.

4 The DTM is a graph that measures the crude birth and death rates over time. This enables us to see the periods when there is a population increase or a population decrease. It sometimes also has another line added that will show the population increase over the same period.

5 The DTM was originally based on the demographic experience that western European countries like Great Britain had gone through. The key triggers, such as the Industrial Revolution and rapid urbanisation that led to economic development, were seen as key triggers for growth. The journey that Great Britain went on clearly exemplified the way that other countries would react. From 1750 to 1800, Great Britain had very high and fluctuating birth and death rates. From 1800 to 1880, the death rates started to fall due to improvements to society that meant people started to live longer. From 1880 to 1950, with the death rate low, the birth rate started to fall due to changes in society, and then finally Great Britain entered its fourth stage where both birth and death rates remained low.

6 Figure 3 shows the epidemiological transition for the USA from 1900. In this case at the early stages in the 20th century, nearly 50% of people still died as a result of infectious diseases with another 15% being subject to cardiovascular disease. This indicates that in the 1900s the USA was still in the age of pestilence and famine. However, things changed quickly as by 1970, the proportion of people dying as a result of infectious diseases had fallen to below 10%. Cardiovascular disease had grown to account for around 55% of deaths and cancer was increasing to over 10%. The USA had moved into the age of degenerative and man-made diseases, showing that people in the USA were living much longer lives.

7 The optimum population is the number of people that, when working with all of the available resources, will return the highest standard of living and quality of life. Overpopulation is when there are too many people for the resources available in an area and underpopulation is when there are too few people for the resources available in an area.

8 The theory of Thomas Malthus was important, as he was the first person to make the link between the growth of the population and food supply. This showed that famines such as that in Ireland was more of a response to a rapidly growing population than just a blight on the potato crop. However, his theory was very simplistic; the balance for sustainability in a population is a very complicated thing, and agriculture and the science and technology used in modern farming is very different today.

On the other hand, Ester Boserup noted that any time that the global population came close to the crisis point, necessity would dictate that humankind would overcome and an innovation would take place that would negate any resource issues. However, even Boserup's theory was seen as too simplistic, as nature is not a closed system but subject to many different influences.

9 China's One-Child Policy has recently been seen as a success as the rapid increase of population was slowed down (some estimates suggest that 400 million babies were prevented from being born between 1979 and 2015). Although the population of China is still growing, it is growing at a slower rate and will, within the next 30 years, start to decline. This has also allowed China's GDP to grow and the GDP per capita to increase rapidly from $310 in 1980 to $7,924 in 2015. However, some of the negative effects are that abortion, sterilisation and female infanticide have all increased. Human rights have been consistently violated in the pursuit of this policy and the country has been left with more boys than girls, an ageing population and families where six adults focus all their attention onto one child, thereby spoiling their 'Little Emperors'.

10

Demographic	Cultural	Physical
■ Population density ■ Population structure ■ Population versus resources (pressure) ■ Migration	■ Nomadic/ sedentary ■ Housing styles ■ Religious practices ■ Fashion/ demand	■ Relief/ landscape ■ Climate ■ Building materials ■ Water courses ■ Natural vegetation

Social	Locational	Economic
■ Type of settlement ■ Political/ social organisation ■ Use of technology	■ Large settlements ■ Water supply ■ Trade routes	■ Types of farming ■ Prosperity/ level of development ■ Level of funding ■ Diversification

11 A green belt will stop urban sprawl as it is a planning mechanism that is used to control and to stop people from being able to plan and build houses in areas of greenfield sites.

12 Suburbanisation is when people who are living in the city (usually the inner city) move out into the suburban areas of the city. Counter-urbanisation, on the other hand, is when people move from both the inner city and the suburbs to places that are further out of the city, towards the surrounding rural towns and into the rural–urban fringe.

13 Urban sprawl can have many different effects on the rural–urban fringe. Most of the effects are not positive — the sprawl will bring much traffic congestion and pollution into the areas. More people wanting to move into rural houses will increase the competition for land and houses and will push house prices up. Green space and animal habitats will come under increased threat. However, on the positive side, the rural–urban fringes may start to develop and improve their services.

14 Urban sprawl can have many different effects on the city centre and inner city. The movement of people out of the inner city frees up valuable space, which can then be redeveloped. People can then build better homes and houses than what might have been available before. However, as people move out, this can also mean that jobs, shops and services may also move out and the city may suffer as a result.

15 Within Northern Ireland there are 9 Areas of Outstanding Natural Beauty, 47 national nature reserves, 43 special areas of conservation and 10 special protection areas, 440 Areas of Special Scientific Interest and no national parks.

16 The main objectives for AONBs in Northern Ireland are to:
- ■ conserve or enhance the natural beauty or amenities of the area
- ■ conserve wildlife, historic objects or natural phenomena within it
- ■ promote its enjoyment by the public
- ■ provide or maintain public access to it.

17

Positive impact of a National Park in Northern Ireland	Negative impact of a National Park in Northern Ireland
■ Would bring an additional £2–4 million in funding. ■ Would create at least 30 new jobs. ■ There would be more opportunities for recreation and increased visitor numbers. ■ There would be increased income from visitors. ■ There would be new housing and planning regulations. ■ There would be increased protection of the landscape.	■ Farmers would have less freedom to choose what crops they farm and what they do with their land. ■ There would be an increase in the number of second homes. ■ There would be an increase in house prices, rates and price of goods. ■ Employment becomes more seasonal and lower paid. ■ There would be an increase in conflict between tourists and locals. ■ High visitor numbers bring problems, including congestion.

18 Redevelopment is when an area is demolished and redesigned. In the inner city this might mean that a street of terraced houses is knocked down and replaced with a block of flats or other cheap housing. Gentrification is when an area is demolished and redesigned. While this attracts new, richer people to the area, the original residents are often forced to move as they cannot afford the new accommodation.

19 The main characteristics of an informal settlement in an LEDC like Nairobi are that:
- – the people usually live in very poor, temporary accommodation
- – slum areas are usually found on poor quality land, in low-lying areas that are at serious risk of flooding/have poor soils, and are close to rubbish tips or steep slopes where landslides are common
- – houses are made of any recycled or reclaimed materials that people can get their hands on;
- – houses are small and lack even the most basic services and
- – houses are often shared with more than one other group or family — 'hot-bedding' is common.

20 Gross national product (per capita) (GNP) is the total economic value of **all** of the goods and services provided in a country through the year (including output generated by the country in other countries), divided by the number of people who live in the country. GDP is the gross domestic product, which is the market value of all the goods and services provided within a country through the year, divided by the number of people who live in the country — it is usually seen as a measure of the standard of living of people in a country.

21 One other economic measure could be the amount of car ownership, which clearly shows the number of people who have enough disposable income to purchase personal vehicles.

22 Life expectancy helps us to understand healthcare as a social measure of development, as the age that someone can live to is often an indicator of the amount of health and social care that someone has access to within a country. If people lack basic medical services they will not be expected to live long but when there are good medical facilities people will generally live longer.

23 The three main indicators within the human development index are life expectancy (a social measure), education (the mean number of years of schooling — a social measure) and gross national income per capita (an economic measure).

24 HDI is a more accurate measure than the PQLI. The PQLI measure uses three measures: infant mortality rate, life expectancy (which both indicate how good a country's healthcare is) and literacy rate (which is a basic measure of education). These are all social measures that require huge amounts of political investment. The HDI, on the other hand, uses one economic indicator, one social indicator and a more robust indicator of education — one that measures the actual access that children have had to education rather than just something that can be achieved in a relatively short time.

25 The eight Millennium Development Goals (MDGs) are:
1. eradicate extreme poverty and hunger
2. achieve universal primary education
3. promote gender equality and empower women
4. reduce child mortality
5. improve maternal health
6. combat HIV/AIDS, malaria and other diseases
7. ensure environmental sustainability
8. develop a global partnership for development.

26 The MDGs have had a huge impact and the UN considered the implementation of the goals as a great stimulus for growth and development among the LEDCs. However, there are still over 1 billion world citizens who continue to live in great poverty.

27 The MDGs were in use from 2000 to 2015 and the SDGs were announced in 2015 and are due to stay in place until 2030. This time, the goals are not just aimed at LEDCs but also at making improvements in MEDCs. The new 'Global Goals' build on the successes and failures of the MDGs, and there are many more targets for countries to aim towards by 2030.

28 Globalisation is a process wherein the world is becoming more interconnected and interdependent. Trade arrangements, goods and services are easily moved around the world. A major indicator for globalisation is the rise of transnational companies, which will dominate and often control the manufacture and trade in an area.

29 Bilateral aid is when aid is given from one country to another. The aid is usually tied so that an MEDC can direct the money towards particular issues and priorities. Multilateral aid is when aid comes from world/international organisations, such as the WHO and the UN.

30 The location of Ghana has played an important part in its development, as it is located on the main sea-based route that Europeans took to get around Africa to Asia. Over time, the location and climate meant that gold and other natural resources could be produced and crops such as the cocoa bean could be planted in this area.

31 An emerging market (or economy) is described as being a country that has some of the characteristics of an MEDC but does not meet all of the criteria. It is a country with an economy that is progressing towards MEDC status. The economy will have a high annual growth rate (GDP), with income between 10% and 75% of the average EU per capita income. Catching-up growth — over the last 10 years a brisk economic growth is noted and growth is evident across a variety of economic sectors.

32 The BRICS countries are Brazil, Russia, India, China and South Africa. These are the five main emerging markets or economies. The group was originally known as BRIC before South Africa joined in 2010. Each of the countries is a newly industrialising country (NIC), but they stand out among other nations as they have large, fast-growing economies and have a growing influence on global politics. MINT countries are identified as Mexico, Indonesia, Nigeria and Turkey. The term was first used by Fidelity Investments but was more commonly used by Jim O'Neill. The term was first used to denote economical and financial growth. Initially the group was to include South Korea instead of Nigeria (MIST countries). However, O'Neill noted that 'Mexico, Indonesia, Nigeria and Turkey all have very favourable demographics for at least the next 20 years, and their economic prospects are interesting.'

Index

Index